A Textbook of Robotics 1
Basic Concepts

A Textbook of Robotics 1
Basic Concepts

Moshe Shoham

Kogan
Page

First published in 1984 by
Eshed Robotec (1982) Ltd,
PO Box 28346,
Tel Aviv 61282,
Israel

First published in Great Britain in 1986 by
Kogan Page Ltd, 120 Pentonville Road, London N1 9JN

British Library Cataloguing in Publication Data

    Shoham, Moshe
      A textbook of robotics.
      1: Basic concepts
      1. Robots, Industrial
      I. Title
      629.8'92   TS191.H

      ISBN 1-85091-143-6 ✓

Printed and bound in Great Britain by
Anchor Brendon Ltd, Tiptree, Essex

Table of Contents

CHAPTER FOUR

CHAPTER FIVE

CHAPTER SIX

CHAPTER SEVEN

CHAPTER EIGHT

CHAPTER NINE

CHAPTER TEN

GLOSSARY

INDEX

Chapter One

Introduction to Robotics

This book presents the basics of an exciting new
field. It is exciting because it extends our
age-old fascination with tools and machines into
experimental areas far beyond mere mechanics and
because of the many ways it promises to affect
our lives. Certainly, it is new, since the
development of robots has taken place only in very
recent times.

The first robot came into use in the early
1960's. By the early 1980's, there were more
than 30,000 robots working in Japan, the
United States, and Western European nations. Even
so, we can safely say that the field is in its
infancy. Forecasts of the future show a dramatic
increase in the rate of robot development, especially
in industry.

Although we marvel at the idea that machines can begin to imitate human actions, even in ways we have not yet thought of, the main motives for the creation of robots have been very practical. First, as modern industry has become more complex, there has been a growing need for getting work done in environments that are dangerous for humans. As an example, work in a nuclear reactor plant often requires contact with radioactive materials.

Second, as robots become more advanced and less expensive, they are being set up in industry situations where working conditions are not so much dangerous as unpleasant for various reasons. These conditions typically involve high degrees of the following:

* Heat

* Noise

* Poisonous gases

* Risk of injury by machines

* Monotonous, boring work

* Extreme physical exertion

Robots have already taken over a number of such unpleasant jobs in industry — welding in automobile factories, for instance, which involves heat, noise, and heavy exertion. The design of a robot for this job is uncomplicated and reasonably priced; robots are obedient, untiring, and precision welders.

It is easy to understand why the robots developed up to this time are applied in industry, rather than in private homes or in service situations like restaurants and shops. The widely-held opinion is that service robots will lag considerably behind industrial robots in the extent of their application. Here are some of the reasons:

* The work carried out by production laborers is generally more dangerous and physically

2

Take Jobs from human

difficult than that done by waiters and *become*
homemakers. *Robot become done on public*

* Production jobs are liable to be more
  monotonous and to require fewer, more
  repetitive jobs. In addition, service jobs
  often involve independent decisions and
  constantly changing schedules. From these
  facts, it is evident that the development of
  an industrial robot is simpler from a
  technological standpoint than that of a
  service or home robot.

* The probability is greater that an industrial
  robot will pay for itself within a reasonable
  time than in the case of a service robot.

* Industrial robots do not normally come into
  close contact with humans, while service
  robots would interact directly with the
  people they were helping, which would create
  a safety hazard.

There is no doubt that science fiction tales about
the amazing feats of robots have stimulated their
development. Visions of their possible applications
have not only excited the general public, but
encouraged many firms to invest large sums in robotics
research and development. Although there are
significant restraints on robot use, as we have noted,
there are two trends over the last twenty years
that make the ongoing evolution of robots almost
certain:

* A constant rise in employee wage levels.

* A remarkable rate of technological advance in
  the field of computers, leading both to
  reductions in robot prices and significant
  improvements in their performance.

It is clear, then, that robotics is not only an
exciting new field, but one of the important directions
in which our era of technology is headed. The purpose
of this book is to explore the fundamentals of

3

robotics, and a logical way to begin is to answer the question:

What is a robot?

That is the subject of the next chapter.

Chapter Two

What is a Robot?

BACKGROUND

In recent years, the subject of new developments in robotics has frequently been in the news – and more than occasionally in the headlines – not only because of innovations in technology, but also because of effects on society.

A number of years ago, the idea was conceived that computers can control mechanical systems in addition to arithmetic operations. This led to the development of <u>computerized numerical control</u>* machine tools, known

* Technical terms will be identified, the first time they occur, by underlining. For later reference, they are collected, with definitions, in a glossary at the end of the book.

5

as <u>CNC</u> machines (see Figure 2-1). These machines are operated, and their speeds are regulated, by computers connected to the machine motors.

In this way, the computer controls the paths of operating tool heads, as well as the final form of the finished products. If we want to change the form of the finished product, all we have to do is exchange the computer program for one that is designed to produce the new product.

FIGURE 2.1: A computerized numerical control machine

The line of thought that brought about the CNC led to
the idea that computers can also control machines
which would be capable of moving in all their <u>axes</u> like
a human hand in space, and perform human tasks. This
new concept was given the name <u>robot</u> – a word meaning
"slave" or "worker" in Czech, which was taken up by the
Czech playwright Karel Capek to refer to machines
designed in man's image to perform all the man's labor.

Now, how has the idea evolved into today's robot?

DEFINITIONS

Robot

> "A robot is a mechanical <u>arm</u>, a <u>manipulator</u>
> designed to perform many different tasks and
> capable of repeated, variable programming.
> To perform its assigned tasks, the robot
> moves parts, objects, tools, and special
> devices by means of preprogrammed motions and
> points."

This definition stresses two main factors:

* The existence of a mechanical arm.

* A flexibility of working capacity,
  enabling the robot to perform various
  operations.

Figure 2-2 shows a machine answering to the definition
of "robot," consisting of a motor-driven mechanical
arm and a <u>brain</u> in the form of a computer that controls
its movement. The computer stores in its <u>memory</u> a
<u>program</u> detailing the course the arm follows. When the
program is run, the computer sends signals activating
the motors which move the arm and the <u>load</u> at the end

7

of the arm, which is held under control by the
<u>end effector</u>.

FIGURE 2.2: A robot and its computer controller

The amount and rate of movement are regulated exactly
by the computer. The program moves the arm through a
series of points. At some of these points, the end
effector is activated to grip or release an object, use
a tool, or do whatever else is required.

The mechanical arm is generally made of metal and is
capable of handling loads of varying weight, depending
on the type of robot. Thus, robots may be divided into
three groups, by load capacity:

* Small (up to 2 pounds)

* Medium (2 to 20 pounds)

* Large (over 20 pounds)

Sensing

We have said that robots resemble humans in certain
ways. Among these is the ability to "sense"
position in the working space, direction and rate
of movement, and various conditions in the environment.
Without his eyes, ears, or other sense organs, man
would be blind, deaf, or otherwise handicapped. The
same is true of robots. To perform the tasks that will
be required of them as their applications expand, they
must be provided with abilities similar to human
senses.

Most of today's robots are "blind," "deaf," and
insensitive in all other ways to the environment. Some
advanced models are equipped with sensors, such as
electronic eyes, but these capabilities are still in a
very elementary stage of development. Not for some
time will robot intelligence permit operations in
unfamiliar environments requiring complex decision-
making.

In addition, most robots now in use in industry are
fixed in one location and are capable of operating
within a strictly defined envelope. They perform tasks
which repeat themselves over and over. Their lack of
sensors means that the components with which they
work must be placed in exactly the right place and
position at exactly the right time.

However, the robot's potential for accuracy of movement
is far greater than that of man. A human being whose
senses are intact can position his hand with an
accuracy of one or two tenths of an inch with his eyes
open, and an inch or two with his eyes shut. By
contrast, robots - even those working without sensors -
can position their "hands" at any given point with an
accuracy in the thousandths of an inch. (In general,
the larger the robot, the less its accuracy). At the
same time, it should be remembered that any change in
the environment or any factors acting on the arm,
including a change in load, will create stress on the
robot arm. Without sensors, it cannot adjust, whereas
the normal human can react very precisely to such
conditions.

Programming

The computer, as we have said, acts as the "brain" of the robot, activating motors which control its moving parts. Robot computers must possess the following characteristics:

* A memory in which programs are stored.

* Connections to the motors' controller.

* Connections to the outside world by means of <u>input</u> and <u>output</u> of information, and to activate operational programs.

* A communication unit enabling control by a human.

Robots are activated by various kinds of computers, from <u>microprocessors</u> to <u>minicomputers</u>. Naturally, more sophisticated robots require more extensive <u>software</u>, or sets of instructions, and larger computers. Some advanced robots require continuous processing and computation, which in turn necessitates the use of high speed computers.

In some robots, the brain is composed of more than one processor. Such a system permits the work to be divided among several processors, each of which has its assigned task, but all of which operate at the same time. For example, some of the processors control the robot arm motors, others handle information from the sensors, and still others <u>interface</u> between motors and sensors.

This division of labor may be compared to a team of laborers headed by a foreman. Each laborer receives an assignment, carries it out, reports to the foreman, and receives a new assignment. By portioning out the work to several laborers, the foreman has time to plan and handle problems that crop up as the work progresses. The fact that a large number of activities are performed simultaneously saves time and enables the team to finish its work more efficiently. Also, each laborer is a specialist. Thus, the foreman does not have to be fully skilled in every task.

TYPES OF AUTOMATION

In industry, the robot is thought of as an integral part of flexible manufacturing systems, in contrast to hard manufacturing systems. What is the difference between hard and flexible automation?

> Machines which are designed to perform specific functions are defined as hard automation. In these systems, every change in standard operation demands a change in machine hardware and setup.
>
> Machines which can be easily programmed – which can change over easily and quickly from one manufacturing setup to another – are defined as flexible automation.

Robots, by contrast with hard automation, are composed of arms capable of executing complex motions and accomplishing many and varied tasks. Hard systems are usually geared for a particular product and are difficult to adapt to another product. Flexible systems, as we have said, are readily convertible to alternate tasks through reprogramming.

Hard Automation

We can illustrate these contrasts. Figure 2–3A shows three fixed mechanical arms along one side of conveyor A. Arm A grips a part taken from the conveyor B and assembles it onto the square part. Arm B grips a part taken from the bowl feeder and assembles it onto the two previously assembled parts. Arm C then transfers the completed assembly to a special area for temporary storage.

Each of the three arms is limited in its motions: an arm can go up and down (along axis 1) or back and forth (along axis 2), but no other motion is possible.

On each of the two axes, only two possible positions exist:

*   Along axis 1, maximum or minimum height.

*   Along axis 2, maximum or minimum extension.

When one of the axes receives a command, it moves until it is stopped by a mechanical end stop. Any change in the assembly operation requires a modification in the machine array, or overall arrangement, that may take considerable time.

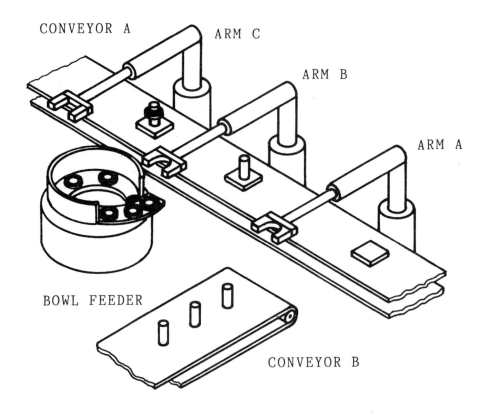

FIGURE 2.3(A): Hard automation: three fixed mechanical arms along a conveyor belt

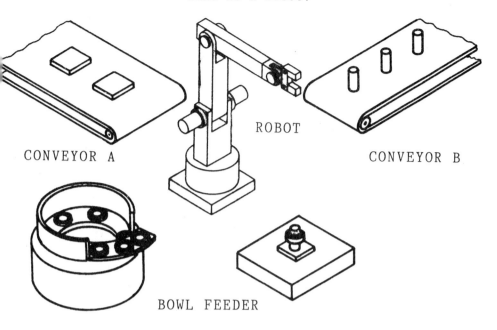

CONVEYOR A

ROBOT

CONVEYOR B

BOWL FEEDER

FIGURE 2.3(B): Flexible automation: a robot performing the tasks of the three arms shown in Figure 2.3(A)

Flexible Automation

Figure 2–3B shows a single robot replacing all three hard mechanical arms, to perform a similar operation. Changes in the assembly operation performed by an array including a robot usually require no more than a modification in the programming.

Note the changed positions of the conveyors, the part feeder, and the area for completed parts. The reason for these changes involves the robot's range of motion. If the conditions for changeover from hard to flexible automation include a requirement that conveyor and feeder positions must not be changed, it is possible to install a mobile robot, mounted on a rail, so that it is capable of reaching the conveyors and the part feeder.

Robots are gaining a place in industries where production consists of small-to-medium runs (thousands to hundreds of thousands of parts). It is still less expensive to use hard automation to produce identical parts in runs of millions, and to use manual labor for production runs in the hundreds and low thousands, but flexible systems are best suited to runs in the middle range.

CLASSIFICATION OF ROBOTS

Naturally, a number of advances have taken place, even in the rather short time that robots have existed. It is possible to classify their evolution by generation, according to their levels of sophistication.

First Generation Robots

Most of the robots in operation today are classified as first generation. These machines are unable to gather any information about their surroundings. They can perform only pre-programmed motions, and the information they can pass back about the operating environment is minimal.

However, first generation robots include a wide range of types, from setups for simple transfer, to advanced models which perform complex tasks, such as painting, and arc welding. The feature they have in common is flexibility, the ability to respond to computer program changes without replacement of hardware.

Second Generation Robots

These models include all the features of first generation robots, plus means for detailed communication with their surroundings. This communication is accomplished by sensing and identification systems, which are discussed in Chapter 9. Such advances necessitate faster computers, with larger memories, and a giant step forward in sensor capability. Development along these lines points, of course, to future generations of robots, which will feature the capacity for independent thought, analysis, and decision-making - in other words, <u>artificial intelligence.</u>

COST LIMITATIONS

We have indicated that the most common use of robots today is in industrial plants, where they replace humans in the performance of difficult, boring, and dangerous tasks. Still, they are in limited use. The main factor preventing the mass adoption of robots is their high price.

The key consideration is how long it takes for a robot to earn back the investment involved in its purchase, installation, and maintenance. This time span is not fixed, but depends on the plant in which the robot is installed and on its application. It is now generally accepted that the minimum is about a year, varying with the following conditions:

* The number of workers replaced by the robot.

* The number of shifts per day it is used.

* Its <u>work rate,</u> or productivity, compared to its cost.

* The cost of required <u>peripheral</u> equipment.

* The costs of engineering and maintenance.

Naturally, the more sophisticated a robot and its software are, the higher its price will be. The main factors determining robot prices are:

* Size.

* Sophistication, or degree of advanced capability.

* Accuracy.

* Reliability.

As a rough guideline, in 1984 the simplest robots cost some $10,000, while large, sophisticated models were priced at over $100,000. Robots equipped with sensors are more expensive. Although robot prices are constantly dropping, it is obvious that costs are still high, and that this tends to hold up mass adoption.

SOCIAL IMPLICATIONS

The adoption of robots in industry has caused alarm in certain quarters. Some employees fear that robots are liable to take over their jobs and put them on the unemployment lines. From a narrow view of the situation these fears are justified. Robots will be assigned to boring, tiring, and dangerous jobs. But this does not mean that humans will be displaced. Far from it.

In the first place, we are still in the early stages of robotics, and numerous problems remain to be solved. As a result, robots skilled and sophisticated enough to replace humans will not emerge for many years.

In the second place, as robots are introduced, those employees who now perform direct production tasks will become supervisors of robots and other machines, or mechanics involved in their maintenance. Their working conditions will improve, and the work itself will be more interesting. In addition, new kinds of jobs will be created as the field advances.

At the beginning of the 1980's, British Prime Minister Margaret Thatcher was quoted as saying: "The English have hundreds of robots and millions of unemployed, but the Japanese have tens of thousands of robots and no unemployment problem." It is possible that, when robots are highly developed and capable of replacing human beings in a significant proportion of production jobs, working hours in production plants will be cut back. At that point, it will be possible to use robots to manufacture the quantities of products required by society at large with decreasing numbers of workers – meaning that humans will be free to do other work, or to pursue their hobbies, for longer periods of time each day.

Is this a desirable kind of progress? No one can be sure. But the age of robotics has begun, and it would be nearly impossible to turn back from it.

Humans versus Robots

Perhaps we can clarify this issue by inspecting it more closely. Where do we stand in relation to these machines, as far as capability is concerned?

The robot has some advantages over humans that come immediately to mind. They include the following:

*   It does not get tired.

*   It does not demand a salary.

*   It can maintain uniform production quality.

*   It does not require such environmental conditions as air conditioning, light, and quiet.

However, to understand our opportunities in the age of robotics, it is helpful to make more subtle comparisons. The table beginning on the following page draws contrasts between human and robot capabilities across a spectrum of factors with which robot designers are constantly concerned. Give it some study.

TABLE 2.1: A comparison of the capabilities of humans and robots

	Robot	Human
Mobility	Generally fixed in one spot, except for a small number of mobile models.	Capable of free motion from place to place.
Arm	Two types of joints: Prismatic and Revolute (defined in Chapter 3).	Revolute joints only.
	Most have only one arm.	Man has two arms.
	Range of motion from several inches to several yards.	Range of motion unlimited.
	Maximum speed of motion ranges from several inches to several yards per second.	Maximum speed of motion about five feet per second.
End Effector	Tool or gripper attached to end of each arm.	Each arm terminates in five-fingered hand, with each finger capable of almost independent motion.
Required Floor Space	Zero space for models suspended from ceiling. Tens of square feet for floor-mounted models.	About ten square feet.

	Robot	**Human**
Load Lifting Capacity	From 1/4 pound for small models to one ton for large models.	Up to about 65 pounds.
Memory	100 to 2,000 motions. Expandable. Absolute, not affected by elapsed time. Can be erased on command.	No evaluation exists. Limited by elapsed time, due to forgetfulness.
Sensors	Current capability mainly for sight and touch. Limited capacity for identification. Decoding of sensed information quite slow. Suitable for a limited range of working conditions. For example, light intensity limits are strictly defined for sight sensors.	Five highly developed senses capable of transferring immense amounts of information to the brain. Decoding quite rapid. Suitable for a wide range of working conditions. Ability to see in bright light and semi-darkness.
Reaction Time to External Signals	Rapid – about ten milliseconds.	Slow – about one second.
Intelligence	Limited. Cannot overcome unforseen situations. Little judgemental ability.	High. Capable of learning from past experience and applying it to unforeseen situation.

	Robot	**Human**
Learning	Tasks are learned once only. No need to refresh memory after long periods without working. Programs can be copied instantly from robot to robot, without relearning. Certainty that all robots of the same type can learn the same task.	Memory must be refreshed after periods without work.Learning must be individual. No certainty that a particular man can learn a particular task.
Consistency	Motions are repeated consistently unless machine is malfunctioning.	Lack of consistency in motions. Can improve with experience, but is subject to psychological influences and fatigue.
Effects of Environment	Little effect. Do not require air to breathe.	Subject to temperature, noise, poisonous gases, other hazardous materials. Requires air to breathe.
Fatigue	None. Periodic maintenance required.	Increases with hours worked, decreases with rest.

What is a robot?

	Robot	Human
Motivation	None required.	Must be interested in his tasks. Work quality subject to psychological influences within and outside the environment.
Absence from Work	May malfunction. Repair time depends on skill of maintenance team.	May be absent due to illness, tea and coffee breaks, strikes, training, personal business.

Chapter Three

The Mechanical Arm

OVERVIEW

In the following sequence of chapters, we will make a more detailed survey of the components of robot operation. This chapter will explore the robot arm, its parts and joints, its work envelopes, and its sources of power. The following chapters will discuss end effectors, methods of controlling the arm, methods of "teaching" the robot what to do, and, finally, how all of these subsystems come together in action.

In the previous chapter, we indicated that the robot is, basically, a mechanical arm, designed to perform motions and to grip objects, controlled by a computer which is informed in various ways about the

environment by sensors. These three main system units
are illustrated in Figure 3–1.

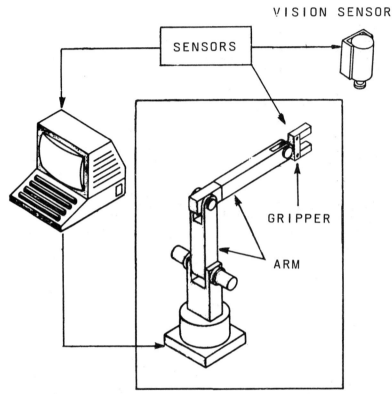

FIGURE 3.1: The three main units of a robotic system
fitted with sensors

In this chapter, as we have said, we want to focus on
the mechanical arm. To repeat ourselves a little, we
will describe it again.

The robot arm performs motions in space. Its function
is to transfer objects and tools from point to point,
as instructed by the controller.

Attached to the end of the arm is an end effector, used
by the robot to carry out its assigned tasks. End
effectors vary according to the job at hand. For
example, in robots used to transfer parts from point to
point, the end effector is a gripper. For operations
such as sanding or welding, the end effector is the
appropriate tool.

> All robot arms, like the human arm, are made up of a series of <u>links</u> and <u>joints</u>. A joint is that part of the arm which connects two links and allows relative motion between them, as shown in Figure 3-2.

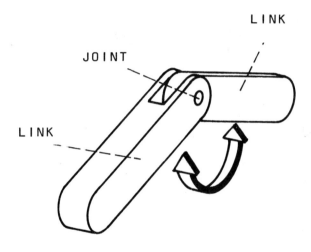

LINK

JOINT

LINK

FIGURE 3.2: A typical joint for a mechanical arm

Each robot has a <u>base</u>, which is normally fixed to the floor, a wall, or the ceiling. The first link of the robot arm is connected to the base, and the last link is attached to the end effector.

In general, the more joints in a robot arm, the greater the dexterity. For example, in Figure 3-3A, note that the robot cannot reach the object hanging behind the wall. However, an additional joint in the robot arm enables it to get around the obstacle, as shown in Figure 3-3B.

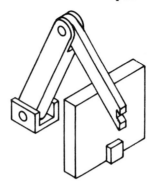

FIGURE 3.3(A): This arm (having two joints) can not be used to pick up the object

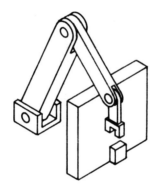

FIGURE 3.3(B): Addition of an extra joint to the arm shown in 3.3(A) allows this arm to be used to pick up the object

TYPES OF JOINTS

Most robot arms include two types of joints, <u>prismatic</u> and <u>revolute,</u> although some include a third type, <u>ball and socket.</u> We will describe them one by one.

Prismatic Joints

This type of connection permits prismatic, or <u>linear</u>, motion between two links. It is composed of two nested links, or of one link sliding along another. In other words, one part can move on a line straight outward or inward in relation to the other part, as shown in Figure 3-4.

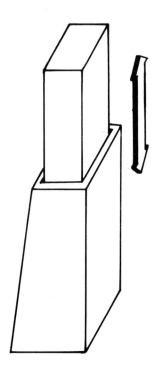

FIGURE 3.4: This machine slide is an example of a prismatic joint; the arrow shows the relative linear motion: this allows sliding but not rotation

The joint between a piston and a cylinder is a prismatic joint, and the motion created by a piston with a cylinder is relative linear motion.

27

Revolute Joints

This connection permits revolute, or <u>rotary</u>, motion between two links. The two links are joined by a common hinge, so that one part can move in a swinging motion in relation to the other part, as shown in Figure 3-5.

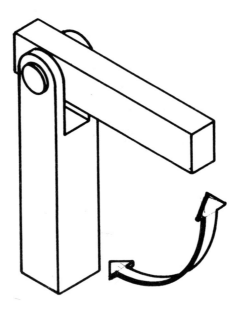

FIGURE 3.5: An arm having a revolute joint; this is a single-axis rotary joint: this allows rotation but not sliding

Revolute joints are used in many tools and devices, such as scissors, windshield wipers, and nutcrackers.

Ball and Socket Joints

This connection functions like a combination of three revolute joints, allowing rotary motion around three axes, as shown in Figure 3–6.

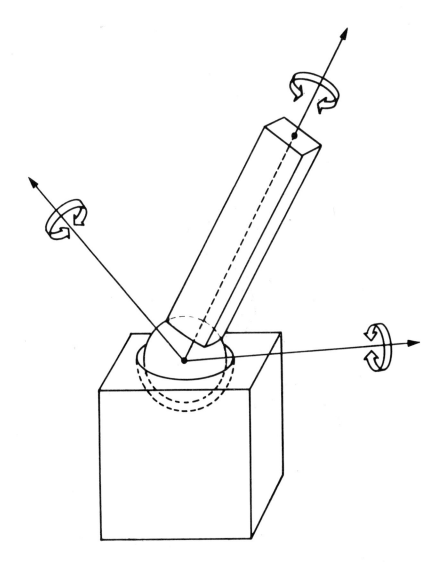

FIGURE 3.6: A spherical or ball and socket joint: this allows rotation around three axes

Ball and socket joints exist in the human body between shoulder and forearm and between pelvis and thigh. They are used in only a very few robots, due to the difficulty of activating joints of this type. However, in order to get the performance of a ball and socket joint, many robots include three separate revolute joints whose axes of motion intersect at one point. Such a joint is illustrated in Figure 3-7.

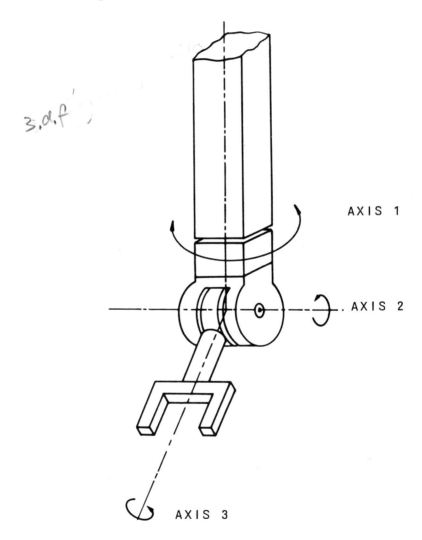

3.d.f

AXIS 1

AXIS 2

AXIS 3

FIGURE 3.7: Three revolute joints with intersecting axes of motion

Degrees of Freedom

The number of joints in a robot arm is also referred to as the number of <u>degrees of freedom</u>. Every joint enables a relative motion between two links and allows the links a degree of freedom. When the relative motion occurs along or around a single axis, the joint has one degree of freedom. When the motion is along or around more than one axis of motion, the joint has two or three degrees of freedom.

Most robots have between four and six degrees of freedom. For the sake of comparison, the human arm, from shoulder to wrist (but not including the joints of the hand) has seven degrees of freedom. This is illustrated in Figure 3–8.

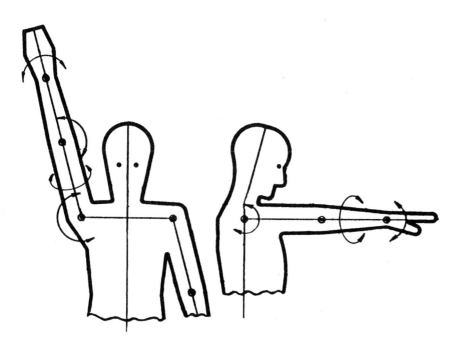

FIGURE 3.8: The seven degrees of freedom associated with the human arm, from shoulder to wrist

31

CLASSIFICATION OF ROBOT BY TYPE OF JOINT

It is customary to classify robots by types of joints – or, more exactly, by the three joints closest to the robot base. This division into classes provides information about robot characteristics in several important categories:

* Envelope (that is, maximum range of motion attainable by the robot in all directions).

* Degree of mechanical rigidity.

* Extent of control over the course of motion.

* Applications suitable or unsuitable to the type of robot.

Robots can be classified by joint into five groups:

* Cartesian

* Cylindrical

* Spherical

* Horizontal articulated

* Vertical articulated

The code used for these classifications consists of a set of three letters, referring to the types of joints (R for "revolute", P for "prismatic") in the order in which they occur, beginning with the joint closest to the base. For example, RPP indicates a robot whose base joint is revolute and whose second and third joints are prismatic.

Cartesian Robots

The arms of these robots have three prismatic joints, and so they are coded PPP. See Figure 3-9.

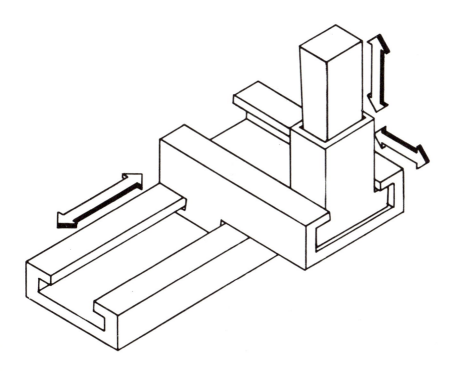

FIGURE 3.9: A typical Cartesian robot linkage system: only prismatic joints are used here

Cartesian robots are characterized by a small work envelope, but have a high degree of mechanical rigidity and are capable of great accuracy in locating the end effector. Therefore, Cartesian robots are suitable for closely calibrated tasks, and machining.

Control of Cartesian robots is simple, due to the linear motion of the links and the fixed inertial load caused by fixed moments of inertia throughout the work envelope.

Cylindrical Robots

These robot arms consist of one revolute joint and two prismatic joints. Their code is RPP. See Figure 3–10.

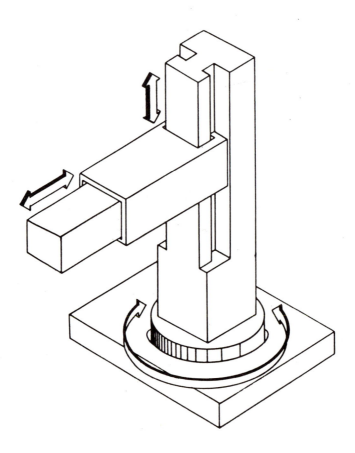

FIGURE 3.10: A typical cylindrical robot; this is best suited for pick and place applications

The work envelope of these robots is larger than that of Cartesian robots, but their mechanical rigidity is slightly lower. Control is a bit more complicated than in Cartesian models, due to the varying moments of inertia at different points in the work envelope, and to the revolute base joint.

Spherical Robots

These robots have two revolute joints and one prismatic joint. Thus, they are coded RRP. See Figure 3-11.

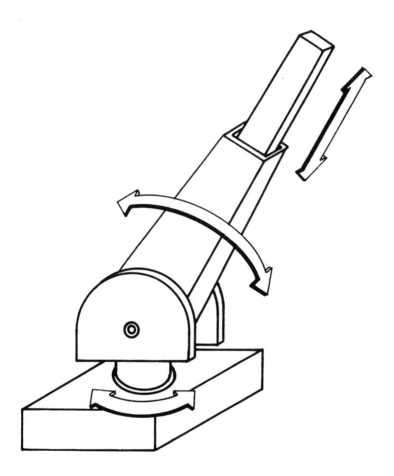

FIGURE 3.11: A typical spherical robot: this is best suited for long, straight reach operations

Spherical robots have a larger work envelope and a lower degree of mechanical rigidity than cylindrical models. Their control is more complicated than in cylindrical robots, because of the rotary motion of the first two joints.

Horizontal Articulated Robots

The arms of these robots have two revolute joints and one prismatic joint, being coded RRP. See Figure 3-12.

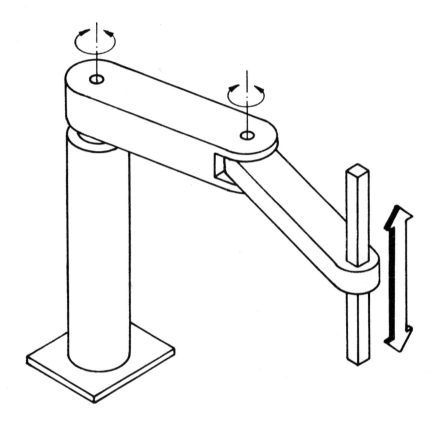

FIGURE 3.12: A horizontally articulated robot; this has similar applications as a cylindrical robot but is more compact

The work envelope of the horizontal articulated robot is smaller than that of the spherical robot, but larger than the Cartesian and cylindrical. They are appropriate for assembly operations, due to the vertical linear motion of the third axis.

Vertical Articulated Robots

This type of robot includes three revolute joints, so it is coded RRR. See Figure 3-13.

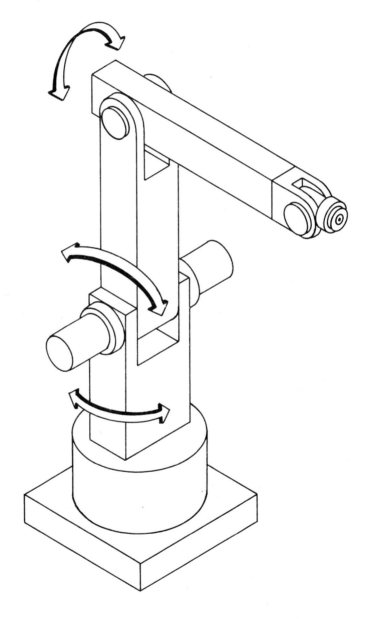

FIGURE 3.13: A vertical articulated robot; this is perhaps the most popular type of robot and avoids prismatic joints

Vertical articulated robots are similar in structure to human arms, which also have only revolute joints. Their work envelope is larger than that of any other type of robot, and their mechanical rigidity is lower. However, no great degree of precision can be achieved in locating the end effector; control is complicated and difficult, because of the three revolute joints and because of variances in the load moment and moments of inertia throughout the work envelope.

COMPARISON OF ROBOT WORK ENVELOPES

In this brief section, we will go through some elementary mathematics that will provide important insight into the calculation of robot capabilities. The comparisons we want to make are illustrated in Figure 3-14, and we will refer to this figure in calculating work envelopes at the end of this section.

Motion of three joints closest to base	Side View	Work Envelope

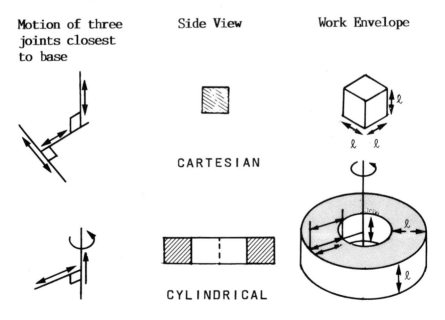

FIGURE 3.14: Work envelopes of various robot configurations

Motion of three joints closest to base	Side View	Work Envelope
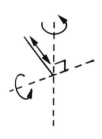	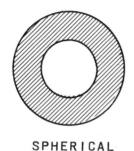	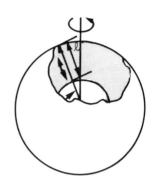

SPHERICAL

		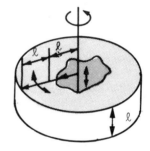

HORIZONTAL ARTICULATED

	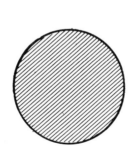	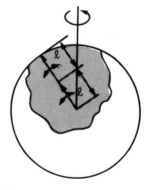

VERTICAL ARTICULATED

FIGURE 3.14 (Continued)
Work envelopes of various robot types

The left-hand column of this figure shows a series of theoretical work envelopes. (In actuality, the work envelope is smaller than in theory, due to mechanical limitations that prevent the end effector from reaching every point in the envelope.)

The middle column shows a schematic diagram of the motion of the three joints closest to the base of each robot type.

The right-hand column shows a side section of the theoretical work envelope.

Now, as a basis for comparison, we have assumed that all the links in all the robots we will evaluate have the same length, which we will indicate by the symbol "ℓ." The unit is equal to one meter. The volume of the work envelope — that is, all the points in space in which the robot can locate its end effector — we will indicate by the symbol V.

Thus, we can describe a cube, a cylinder, and a sphere by the following formulas:

*   Cube   : V = $\ell \times \ell \times \ell = \ell^3$

*   Cylinder : V = $\Pi \ell^2 \times \ell = \Pi \ell 3$

*   Sphere   : V = $4/3 \Pi \ell^3$

With these symbols and formulas, we can calculate the volume of the theoretical work envelope for each robot which we earlier classified by type of joint.

Cartesian Robots

The robots are capable of reaching any point in a cube of edge length ℓ.

Therefore, the formula for the work envelope volume in this type of machine is:

$$V_{Cartesian} = \ell^3 \ (meter)^3$$

Cylindrical Robots

These models can reach any point in a cylinder of height ℓ and radius 2ℓ , except for the points in an inner cylinder of height ℓ and radius ℓ . So the formula for the work envelope is:

$$V_{Cylindrical} = \ell \ \{\Pi(2\ell)^2 - \Pi\ell^2\} = 3\Pi\ell^3 = 9.42\ell^3$$

Note:
In most of the cylindrical robots the work envelope is limited to the area close to the axis of rotation.

Spherical Robots

These robots can reach any point in a sphere of radius 2ℓ, except for the points in an inner sphere of radius ℓ. The formula for the work envelope volume is:

$$V_{Spherical} = \frac{4\Pi}{3}(2\ell)^3 - \frac{4\Pi}{3}\ell^3 = \frac{28\Pi}{3} \ \ell^3 = 29.32\ell^3$$

Horizontal Articulated Robots

These robots are capable of reaching any point in a cylinder of height ℓ and radius 2ℓ. The work envelope volume is:

$$V_{Horiz. \ art.} = \ell \times \Pi(2\ell)^2 = 4\Pi\ell^3 = 12.56\ell^3$$

Vertical Articulated Robots

These models can reach any point in a sphere of radius 2ℓ. So the formula for the work envelope volume is:

$$V_{\text{Vert. art.}} = \frac{4\Pi}{3}(2\ell)^3 = \frac{32\Pi}{3}\ell^3 = 33.51\ell^3$$

We have said that these robots have a progressively greater work envelope as we go from Cartesian through the list to vertical articulated. The ratio between the extreme-case work envelopes is impressive:

$$\frac{V_{\text{Vert. art.}}}{V_{\text{Cartesian}}} = 33.51$$

That is to say, the theoretical work envelope of a vertical articulated robot, with two links of length ℓ is 33.51 times greater than the theoretical envelope of a Cartesian type with three links of length ℓ.

SUITABILITY FOR PARTICULAR TASKS

Now, continuing our discussion of mechanical arms, we can take a slightly different view of the work capability of various types of equipment.

Examination of the types of joints and their order (collectively referred to as the kinematic configuration) permits the designer to evaluate a robot's work envelope, mechanical rigidity, and ease of arm control. As a result, it is possible to estimate which tasks will be most suitable for each type of robot.

As previously stated, robots are classified according to the three joints closest to the base. The motion of these three joints enables the robot to move its effector to any point within its work envelope, but without the ability to control the <u>orientation</u> of the end effector in space.

In most tasks, the orientation in space between the end effector and the workpiece is important; it is not enough for the robot to reach the workpiece, but it must arrive with the end effector in a certain attitude in relation to the workpiece. This capability can be achieved by adding joints to the wrist of the arm, so that it has more degrees of freedom. The wrist may include as many as three joints, which enable it to perform the following motions:

*   <u>Pitch,</u> which is movement of the end effector up or down.

*   <u>Roll,</u> which is rotation of the end effector clockwise and counterclockwise.

*   <u>Yaw,</u> which is movement of the end effector to the right or left.

We can use the arm in Figure 3-7, back in the section on ball and socket joints, as an example. In this case, axis 2 permits up and down "hand" motion (pitch), while axis 3 allows clockwise-counterclockwise motion (roll). Rotating axis 1 to a new position 90 degrees from the current position changes the motion of axis 2; instead of pitch motion, it now executes yaw motion.

Figure 3-7 is an example of many possible robot <u>wrist</u> joint arrangements. Obviously, the ability to <u>alter</u> wrist attitude in various ways is a vital concern in robot design.

Figure 3-15 is another example. It shows the initial position of the mechanical arm and two possible changes in the location of the end effector or in its orientation. Alternate position 1 shows a change in end effector orientation without location change. Alternate position 2 shows a change in location without orientation change.

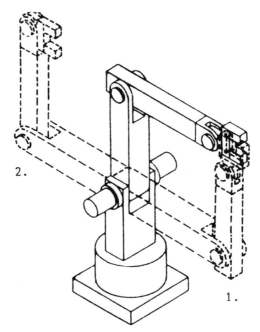

FIGURE 3.15: Movement of a mechanical arm: (1)
involving a change in gripper orientation without
involving a location change, and (2) involving a
gripper location without orientation change

LINK CONSTRUCTION

Another important factor in robot construction is the
load incurred by the machine's own moving parts. The
mechanical arm must be light in weight, but have a high
degree of rigidity. A heavy arm necessitates large
motors, making the robot much more expensive. An arm
with low rigidity reduces the robot's precision, due to
vibrations and response to stress.

An example of an arm with a low degree of rigidity is a
fishing rod. The precision with which the end effector
(the hook) can be located is extremely low, measurable

in yards. Most robots require a precision measurable in fractions of an inch.

To increase rigidity in the mechanical arm without increasing its weight, structural forms like the hollow shell are often used. An arm of shell construction has a better stiffness to weight ratio than an arm of solid construction, as shown in Figure 3–16, where reactions to identical loads are compared.

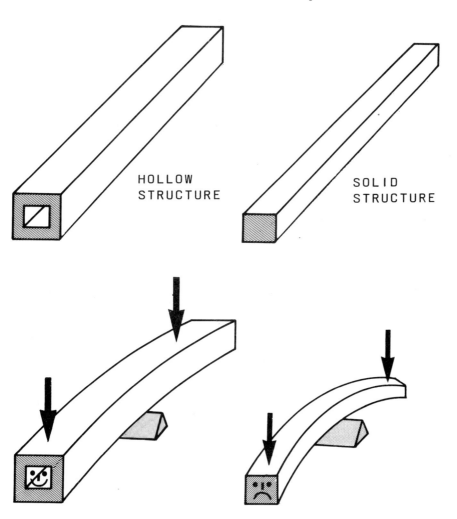

FIGURE 3.16: Reactions to identical loads

In spite of attempts to reduce the weight of mechanical arms, they are still very heavy in relation to the payloads they can move. At present, the ratio between payload and arm weight is about 1:20; that is, a robot which weighs twenty pounds can move a load of up to one pound. By comparison, the ratio between the weight of a human arm and the payload it can move is of the order of 1:1.

ROBOT ARM DRIVE UNITS

Now we turn to a very different aspect of robot arm construction – the sources of the moving power.

The muscles of the human body move the various parts of the arm. Robots, too, must have drive units that move the parts of the mechanical arm. Various types of drives exist, and they are generally classified as follows:

* Revolute and prismatic

* Electric, hydraulic, and pneumatic

* Direct and indirect

Revolute and Prismatic Drive

Revolute, or rotary, drive is a motor. When it is connected to its energy source (electricity, oil pressure, or air pressure), the motor axis responds in a rotary motion. The load attached to the motor axis is moved by the rotation of the axis.

Prismatic drive is a hydraulic or pneumatic cylinder, making up a prismatic joint. Linear motion may result from the rotary motion, through the use of a lead screw, or an arrangement of rods pushed by the motor, which convert the rotary motion into linear motion.

Electric, Hydraulic, and Pneumatic Drive

A common method of robot classification is by source of drive enrgy. The main types of drive, according to energy source, are electric, hydraulic, and pneumatic.

Electric Drive

This type of drive is accomplished in robots by means of electric motors connected to a source of electric current. Various types of electric motors are used as robot drives: DC motors, stepper motors (which are a sub-class of DC motors), and AC motors.

Most new robots are DC motor driven, rather than hydraulic and pneumatic, mainly because of the high degree of precision and the simplicity of the controls in electric motors.

Following are the main advantages of electric drive:

* Allows efficient, precise control.

* Involves simple structure and easy maintenance.

* Does not require an expensive energy source.

* Costs relatively little.

Following are the main disadvantages of electric drive:

* Is unable to maintain a constant moment at varying speeds of revolution.

* Is liable to damage from loads heavy enough to stop the motor.

* Has a low ratio of motor output power to motor weight, requiring a large motor in the arm.

47

Hydraulic Drive

These units, when used in robots, include the following:

*   Motors for revolute motion

*   Cylinders for prismatic motion

Hydraulic drive units cause motion in parts such as pistons, by compressing oil. Figure 3-17 shows a prismatic hydraulic unit. Control in this unit is accomplished by a controller that activates the valves. This causes the piston to move, creating motion of the load, because of differences in oil pressure between the two portions of the cylinder which are divided by the piston.

Following are the main advantages of hydraulic drive:

*   Maintains high, constant moment over a wide range of speeds. Moment remains high even when starting (zero velocity), and provides good carrying ability.

*   Enables precision of operation – somewhat less than electric drive, but much more than pneumatic drive. This is true because oil, unlike air, is not compressible. That is, it does not change its volume as a result of changes in pressure.

*   Can maintain high moment over long period of time without damage when stopped.

The main disadvantages of hydraulic drive are:

*   Requires an expensive energy source.

*   Requires extensive, expensive maintenance.

*   Requires expensive precision valves.

*   Is subject to oil leaks from the system, creating a nuisance.

The mechanical arm

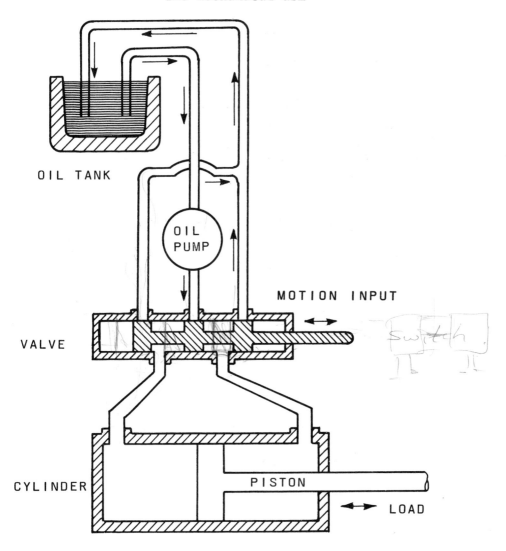

FIGURE 3.17: A prismatic hydraulic drive unit

Pneumatic Drive

These units are similar in structure to hydraulic drives. They include:

*   Pneumatic motors for revolute motion

*   Pneumatic cylinders for prismatic motion

Figure 3–17, which shows a hydraulic cylinder, is quite similar to a pneumatic cylinder, but with one basic difference – air, not oil, flows through the system, moving the piston and creating motion of the load.

Pneumatic drive units are often used in simple automatic systems, such as those shown in Figure 2–3A. These systems are machines in which the piston moves until stopped by a mechanical end stop, thus achieving a high degree of precision in stopping.

Today, by contrast to the widespread use of pneumatic drive in simple automatic systems, the number of robots using pneumatic drive is quite small. This is because the high compressibility of air reduces their ability to achieve precision control over a continuous area.

Pneumatic drive is applied on a large scale to gripper motion. Many electrically and hydraulically driven robots have pneumatic grippers. Often, there is no need for controlled gripper operation; that is, the gripper can be either fully open or fully closed. These robots can use pneumatic gripper drive effectively.

Following are the main advantages of pneumatic drive:

* Enables extremely high operating velocity.

* Costs relatively little.

* Is easy to maintain.

* Can maintain a constant moment (lower than that of hydraulic drive) over a wide range of velocities.

* Can maintain high moment over long periods of time without damage when stopped.

Following are the main disadvantages of pneumatic drive:

* Cannot achieve high precision.

* Is subject to momentary arm vibration when the pneumatic motor or cylinder is stopped.

This list of disadvantages appears short. However, the inability to achieve high degrees of precision and the problem of vibrations are serious enough to rule out pneumatic drive in most industrial robots.

To summarize, then, electric drive is best in applications involving:

* High precision of position

* Transfer of small and medium-size loads

* Too little room for oil and air compressor systems

Hydraulic drive works best in situations involving:

* Transfer of large loads, even to 2,000 pounds or over

* Medium to high precision in location and velocity

Pneumatic drive is preferred in applications involving:

* Low precision

* The necessity for low cost

* Extremely high velocities

* Transfer of small and medium-size loads

Direct and Indirect Drive

Drive units like motors exert forces on the robot joints and cause motion of each robot link. The combined motion of the links creates the arm motion.

Direct drive means that the motor is mounted directly on the joint it moves. If the motor is mounted far from the joint – usually near the robot base – the drive is indirect. Motion in indirect drive is transferred from the motor to the joint by <u>transmission</u> devices which include such elements as rods, chains, and gears. Figure 3-18 shows direct and indirect drives in a typical robot.

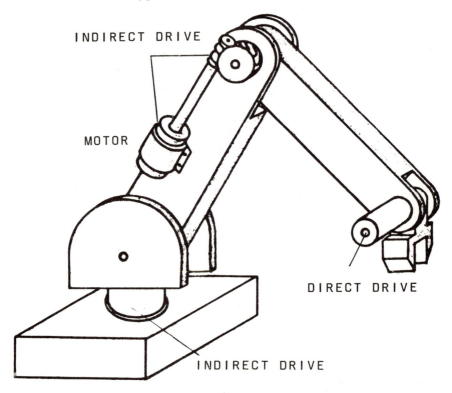

FIGURE 3.18: A robot may have direct and indirect drive systems

Following are the main reasons why indirect drive is preferable to direct drive:

* Reduces the weight of the mechanical arm by mounting motors at the robot base.

* Enables change in the revolute velocity of the joint as a proportion of that of the motor.

The mechanical arm

Following are the main disadvantages of indirect as opposed to direct drive:

* Lacks precision of joint operation, due to the <u>mechanical freedom</u> at the connecting points between the motion transfer devices. Mechanical freedom is a looseness between tooth wheels, gears, belts, and other transmission devices. It is also called <u>backlash</u>.

* Loses considerable power because of the need to move and turn the motion <u>transfer</u> devices located between the motor and the joint.

With these observations about robot drive units, we have completed our discussion of the mechanical arm, for the time being. We will now take up the next topic in our sequence on the components of the robot – the controller.

Chapter Four

The Robot Controller

The controller is that part of the robot which operates the mechanical arm and maintains contact with its environment. The device is a computer composed of <u>hardware</u> and software, combined to enable it to carry out its assigned tasks.

How does the controller execute the process of control, and what are the various methods used in this process?

There are a number of levels in robot control, and they are the first topic of this chapter.

ROBOT CONTROL LEVELS

Robot control may be divided into three levels which make up the control hierarchy. Each level is higher than the previous one. The levels are as follows:

* Actuator control, or the control of each robot axis on a separate basis. This is the lowest level of control. (Level 1.)

* Path control, or the control of the robot arm with coordination between the axes to form the required path. This is the intermediate level of control. (Level 2.)

* Control of the motion of the arm in coordination with the environment. This is the highest level of control. (Level 3.)

These three levels of control exist in the human body, as well. Figure 4-1 shows the relationship between the various levels in performing a task – in this case, lifting an object.

As we have indicated before, the third level of control we have just listed – coordination with the environmnet – is still in the experimental stages. We will discuss this topic in Chapter 9. Here, we will limit ourselves to the first two levels – actuator control and path control.

Actuator Control

Actuators are the units which cause motion of the robot axes. Each axis of motion of the robot arm includes, at least, a joint, a link, and an actuator. In some robots, the axes also include motion transfer devices, as well as units to identify the relative position of the links. An axis including a motion transfer device is said to have indirect drive. An axis including a unit to identify relative link positioning is said to have closed-loop control, which will be explained in the next section of this discussion.

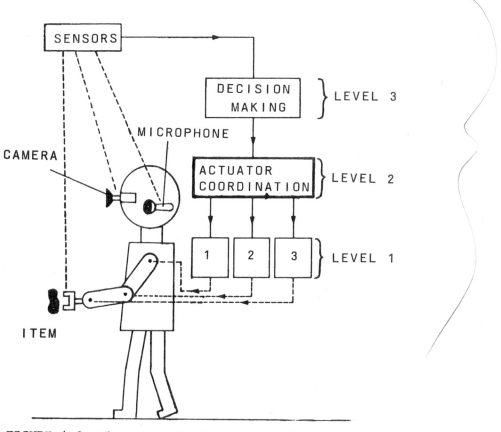

FIGURE 4.1: The control hierarchy and the relationships between the levels of control involved in the movement of an object

The motor is activated by an electrical signal received from the computer. The fact that a great deal of power is needed to operate the drive units means that the computer, itself, is unable to supply this power; therefore, a driver, or amplifier, is used to increase the strength of the computer signal.

The computer puts out digital signals; most of the robot actuators, on the the other hand, operate analog current/voltage levels. A digital/analog converter is required to interface between the computer signals and the drivers. Figure 4–2 presents a schematic diagram of the array used to activate an electric motor.

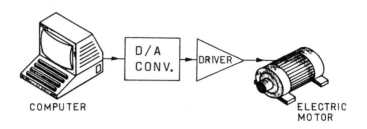

FIGURE 4.2: The computer controls the electric motor via a digital analog converter and driver system

Converters and drivers are also used in hydraulic and pneumatic drive systems. The difference is that, in these types of systems, the electrical unit moves only the valve (see Figure 3–17), while the energy for the actuator is supplied by an external unit, such as an air or oil pump.

Closed-loop Control

We now have an idea of how the first level of control works. Before we move into the next level, path control, we want to introduce the concept of feedback, which is achieved through closed-loop control.

The principle is shown in Figure 4–3, which represents a man looking for his place in a row of seats.

The commands to the man's feet – whether to move forward or backward or stop – are given by the brain as information is transferred to the brain by the eyes. The eyes provide a flow of data on the man's present location; the brain compares this information with the number printed on the ticket and makes any necessary commands. As long as the two are not identical, the feet will continue to move, and will only stop when the man is standing in front of a seat whose number matches

58

the one on the ticket. This is an example of closed-loop control.

What, then, is an <u>open-loop system?</u>

FIGURE 4.3: The principle of closed-loop control

Imagine that the man in Figure 4-3 is blind. He knows he has to reach seat 17. He may count steps, assuming he hnows the number he has to take, and when he has reached what he thinks must be seat 17, will sit down. But will it, in fact, be seat 17? He cannot be sure. The blind man is working in an open-loop system; he receives no feedback to allow him to correct any errors he may have made.

The principle of closed-loop control is applied in nearly all existing industrial robots. Very few work in an open-loop mode.

Each joint, in closed-loop control, includes a feedback unit known as an <u>encoder,</u> which provides information on the relative positions of the links. Whenever a

condition occurs that interferes with programmed motion
– such as a change in the load on the arm – the encoder
signals the controller. The controller compares this
information on the actual status of the joint with the
command originally sent to the actuator in that joint.
The difference between programmed and actual joint
status is called the error signal. The controller will
then take measures to reduce the error to zero, by
instructing the actuator to move the joint until the
desired status is reached.

The simultaneous performance of this type of action on
all actuators is the very heart of low-level actuator
control. A schematic diagram of this control operation
is shown in Figure 4-4. The comparator gives the error
signal, causing the system to cancel the error status
by correcting the joint motion.

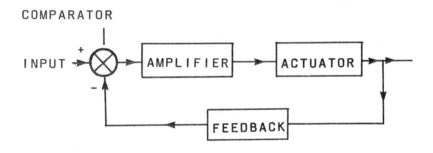

FIGURE 4.4: A typical feedback system used in
low-level actuator control

In order to obtain a high degree of precision in robot
performance, closed-loop control must be used
regardless of the type of robot drive in use
(electric/hydraulic/pneumatic; direct/indirect;
prismatic/revolute) and of the type of joint
(Cartesian, cylindrical, and so on). Figure 4-5 shows
a schematic diagram of closed-loop control in a
prismatic hydraulic (or pneumatic) drive unit. Compare
this diagram to Figure 3-17, which does not include
closed-loop control.

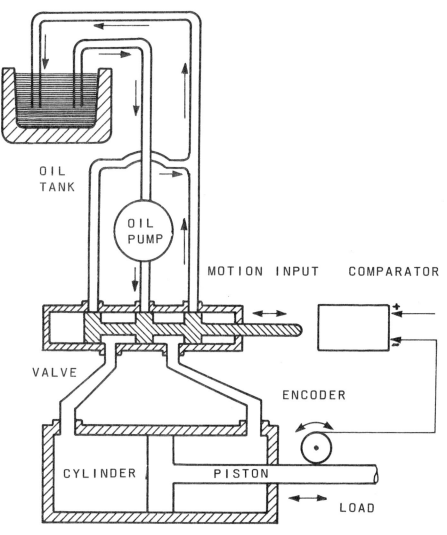

FIGURE 4.5: Closed-loop control of a prismatic hydraulic drive unit

The encoder is connected to the piston, and measures the linear movement of the piston. This information is transferred to the valve control unit, which immediately acts to reduce the error to zero.

Now, how do encoders work?

Most encoders are mounted on the robot arm and are connected to the robot axes of motion, either directly or indirectly. A number of types exist.

Encoders and Tachometers

The most common is the <u>optical rotary encoder.</u> Its principles of operation are shown in Figure 4-6.

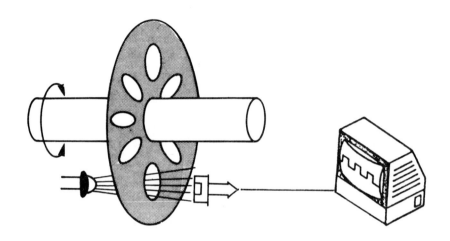

FIGURE 4.6: A rotary optical encoder

This encoder is made up of three essential parts:

* A light source (such as a light-emitting diode).

* A light detector.

* A slitted disc, which revolves between the light source and the light detector.

The robot controller

The disc is directly or indirectly connected to the shaft actuating the robot. It causes the detector to pick up, at intervals, the light transmitted from the light source. The detector creates a corresponding digital electrical signal, which can be seen in Figure 4-6. This signal is made up of a continuous series of low- and high-level signals; a low-level signal appears when the detector "sees" the light source, and a high-level signal appears when the disc blocks the light source so that the detector does not "see" it.

A high-level signal followed by a low-level signal is called an electrical pulse. The number of electrical pulses created by the detector is proportional to the extent of movement and/or to the actual angle of rotation of the robot axis.

A high degree of precision in robot control means the ability to observe small changes in angle or in extent of movement. In order to increase the degree of precision of the revolving optical encoder, we can increase the number of slits spaced around the disc.

The encoder we have just described supplies information on the actual robot position. However, for efficient system control, this information is not enough. The velocity of the various robot axes must also be known, in order to prevent fluctuations in robot motion.

Consider a situation in which all the robot joints are in the positions programmed by the controller. If a deviation occurs, at this point, in the velocity of one or more of the robot axes, we will be able to locate the problem only when it causes a deviation in the robot positioning. In this situation, the closed-loop control will operate to reduce the deviation to zero. However, by that time, unnecessary fluctuations will have occurred. Velocity control serves to prevent such deviations. Identification of velocity errors prevents deviations, and robot motion is smooth and free of fluctuations. The motion is damped.

A component used for velocity control is a <u>tachometer</u>. Figure 4-7 shows a unit which includes a motor, an encoder, and a tachometer on a common shaft.

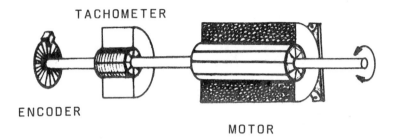

TACHOMETER

ENCODER

MOTOR

FIGURE 4.7: An encoder and tachometer unit is used to control the speed of the motor

Coordinated Control

Up to now, we have discussed closed-loop control on individual actuators. However, robot motion is based on coordinated motion of all robot actuators. The conditions in which the actuators operate are not exactly the same. Their loads are different, as are their velocities, moments of inertia, and so on. The varying conditions may require a different kind of planning for the control loop. For example, see Figure 4-4, which involves a change in gain of the amplifier at the input to the electric motor (or at the input to the hydraulic actuator valve).

In advanced robots, each of the actuator control loops is controlled by a microcomputer, which sends the desired actuated signal to the motor and receives feedback on link and joint positions from the encoders. These microcomputers are, in turn, controlled by another microcomputer, which executes control on a higher level – coordination of all joint operations.

If we want to move the end effector to a certain point, we can dictate the coordinates of that point to the controlling computer, which will then coordinate the motion of the various joints so that, when motion is concluded, the end effector will be located at the desired point. The human operating the robot does not have to control each axis separately, nor to coordinate the motion of the various axes. That is the job of the controlling computer, which will be discussed more fully in the next section.

Path Control

We said earlier that actuator control is the lowest level of control. We noted at that time that the intermediate level is path control. We will take up that topic now.

Any task executed by a robot can be considered as a series of operations, by means of which the end effector is moved by the robot arm between given points and operated as programmed at those points.

We have said that, in order to bring the end effector to a given point, the robot joints must be made to move in coordination. Figure 4-8 shows the motion of a robot from point A to point B. The arrows in the figure show the angular motion of each of the three joints closest to the robot base.

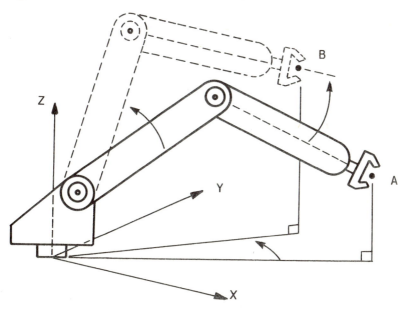

FIGURE 4.8: Movement of a robot arm; the initial and final positions of the joints are shown

65

The coordination of these motions – or path control – can be divided into two methods: <u>point to point control</u> and <u>continuous control</u>. Although the robot moves from one point to another in both methods, and at some of these points performs a task, the methods differ in the manner in which the robot arm moves between these points, as we will see.

Before we begin a discussion of each method, however, we want to define the following terms:

* <u>Point</u>: a location in space towards or through which the end effector is moved by an operation of the robot arm. The arm may stop at a point in order to activate the end effector, or it may continue immediately to another point in the path.

* <u>Step</u>: a part of the robot's operating program. At each step, the robot carries out an activity. For example, the robot may move the end effector from one point to another; or may activate the end effector at a certain point to carry out a task; or may wait at a certain point to receive information from sensors which will instruct it how to carry on.

* <u>Series</u>: a collection of steps which combine to form the robot's operating program.

Point to Point Control

In this type of control, we first define a collection of points for the robot. We then build a series and store it in the memory of the controller. When we run the series, the robot arm will move through the various points according to the order of steps in the series.

At any step calling for point-to-point motion, the arm controller knows the extent of motion required of each robot axis, and sends commands to the motors. At the end of the step, the end effector will be located at the desired point. However, during the time it takes to complete the step, we do not know the path taken from point to point.

All actuators taking part in a given step begin to operate at the same time. But they do not necessarily stop at the same time. In less advanced robots, the actuators will complete their operation at different times. As a result, arm motion will be jerky, with sharp, angular changes in direction. In advanced robots, the controller regulates motion so that the actuators begin and end at the same instant. Thus, motion is smoother and, in short motion segments, resembles a straight line. This type of motion regulation is known as interpolation.

Note that, in both of these instances, the control is of the point-to-point type, since the location of the end effector is specified at the end of the path, but the path between points is not dictated.

Point-to-point control robots are generally used in series where the end effector is not required to perform work while the arm is in motion. More than half of the world's robots operate by point-to-point control. Typical applications of this type of control are spot welding and loading and unloading of machines.

Continuous Path Control

We will have more to say about point-to-point control later, but we will turn now to a description of continuous path control.

This method is more complicated and more expansive than point-to-point control because the arm must move along an exactly defined path. Actuator motions are coordinated by the arm controller at every instant, so that the actual path will resemble the programmed course as closely as possible.

The robot path may be defined by either of two methods, which we designate here as method A and method B.

Method A.
In this technique, the robot arm is moved manually along the desired path, while the controller records in its memory the positions of the joints at every instant from information supplied by the encoders. When the series is run later, the controller sends commands to the actuators according to the information in its memory. The arm thus repeats the path of motion precisely.

Method B.
In this technique, the path is defined by a desired course of motion, such as a straight line or an arc passing through given points. The controller calculates and coordinates the motion at every instant.

Continuous path control may be thought of a point-to-point control with a large number of points distributed at close intervals. For this reason, continuous path control requires a controller with a large memory and/or the ability to make fast calculations of interpolation.

Continuous path control robots are used in series where the end effector must perform work while the arm is moving. Some typical applications are spray painting and arc welding. Further applications will be discussed in Chapter 8, but it is useful to note here that continuous path control is essential in robot systems involving sensors, where arm motions are not programmed in advance, but are executed according to sensor information, in real time.

Figure 4-9 illustrates the two methods of control: point-to-point control and continuous path control. In this figure, the end effector of a two-joint robot arm (one revolute and one prismatic) is moved from point A to point B, located at an equal distance - designated by R - from the robot base.

Point-to-point control is shown in Figure 4-9A. The arm controller instructs the revolute joint to move at an angle designated by Θ, and does not send any instruction to the prismatic joint. The path of the end effector, then, is an arc with radius R. The Y

68

coordinate of the end effector changes at every point along the path.

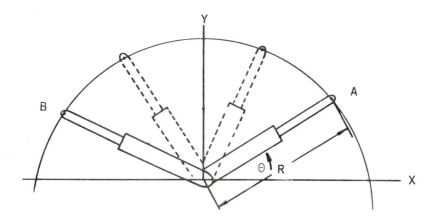

FIGURE 4.9(A): Point-to-point control

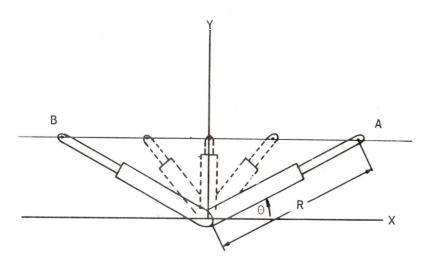

FIGURE 4.9(B): Continuous path control

Figure 4-9B shows continuous path control, whose aim is to move the end effector from point A to point B in a straight line. The arm controller instructs both joints to keep moving simultaneously in such a way as to keep the Y coordinate of the end effector constant. The extent of motion of both joints (angular motion of the revolute joint, linear motion of the prismatic joint) must remain precise throughout the entire course of motion, in order not to deviate from a straight line.

Figure 4-10 shows the changes in arm extension (4-10A) and an angle between the arm and the Y axis (4-10B) for both methods of control.

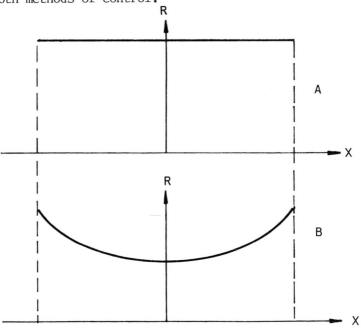

FIGURE 4.10: Changes in arm extension(A) and Θ(B) during point-to-point and continuous path control

OFF-LINE PROGRAMMING CONTROL AND REAL-TIME CONTROL

A principal difference between various controllers relates to the time in which the path of motion is calculated, and to the ability to make changes in the path while the arm is in motion. This section will explain the nature of the various methods and the differences between them.

As it learns the path of motion, the arm controller receives numerous commands from the human operator, such as: "Move the robot to point X." The controller processes these commands, turning them into required motions of the robot joints in terms of angles and of linear lengths. The contrast between off-line programming and real-time control is in the mode of directing the path.

Off-Line Programming Control

In this mode, the robot controller stores the path of motion in its memory as a series of points and corresponding motions of the various joints.

While the robot's operating program is being run, the controller using off-line programming control does not have to do path computations. Instead, the controller gets movement commands from the memory which have already been processed. Controllers using this method of control cannot perform computations while the robot's operating program is running. Therefore, they cannot handle series where changes occur during the run - such as a series involving the use of sensors.

Off-line programming control does not require fast, complex computers. For this reason, it is less expensive than real-time control.

Real-time Control

In this mode, the controller receives general instructions as to the path of motion. While the arm is in motion, the controller must calculate the extent of motion of various joints in order to move along the required path. Information received from the sensors regarding changes in the robot's environment while the arm is in motion is processed by the controller in real time. ,

Real-time control is preferable to off-line programming control, since it is more flexible in its ability to change the path of action while a task is being carried out. This flexibility requires a more complex controller, including a computer fast enough to process the information without slowing robot operation.

Figure 4-11A shows a robot moving from point A to point F along a straight-line path. Figure 4-11B demonstrates the ability, under real-time control, to change the path when point G replaces point F as the target point.

In Figure 4-11A, which describes off-line programming, the path of motion is fixed; it cannot be changed while the arm is in motion. The controller reads the commands for creating a straight-line path from its memory. These commands move the arm from point A to point B, from point B to point C, and so on, until point F is reached.

In Figure 4-11B, which describes real-time control, the controller reads from its memory points A and F only, then calculates the intermediate points required to form a straight-line path while the arm is in motion. If the controller receives an instruction from a sensor to change the target position to point G, it can make calculations while the arm is in motion and instruct the arm to this new target point.

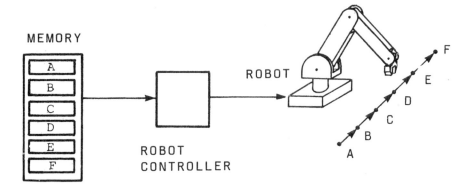

FIGURE 4.11(A): Off-line programming control

72

With these concepts as background, we can take the next step in our discussion. During the course of our exploration of robotics systems, we have said a lot about programming. As a final section of this chapter, we will look into that important topic now.

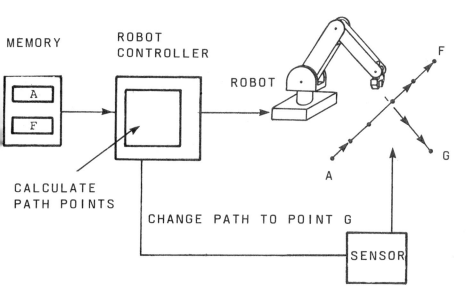

FIGURE 4.11(B): Real-time control

CONTROLLER PROGRAMMING

As we have already noted, industrial robots are highly flexible machines, both in operation and in ability to handle changes in their assigned tasks. The degree of flexibility of any robot depends on its controller – which means, primarily, on controller software.

It is important to differentiate between two components of this software – <u>user program</u> and <u>controller software</u>.

73

User Program

This component is written by the robot operator for each series performed by the robot. It consists of the collection of points along the path and of the operations performed at these points by the end effector. Program commands are written in <u>high level language</u> - for example: "Go to point 4 and open gripper."

Writing a user program for a robot does not require a computer expert. Any employee can learn to write a program with a few day's instruction. During the time the robot remains in use in a plant, many user programs will be written for different production processes. All of them can be retained in the controller memory. The time varies from a few minutes to a few hours, or few days depending on the complexity of the operation.

Controller Software

The second component of robot software is written by the manufacturer. It is responsible for processing the user program commands and converting them into commands for the robot.

Controller software must be written by an experienced software expert with detailed knowledge of the controller computer. It may take up to several years to write, as the robot is being developed in the factory.

The robot's degree of sophistication and its ability to perform varied tasks are determined by the controller software. As we have indicated, advanced robots can coordinate joint motions, execute paths according to functions, and react in real time to events in their environment.

Controller software in advanced robots is complex. It includes hundreds of computations which must be performed very rapidly while the robot is in motion. It becomes more complex as the number of degrees of freedom increases.

For example, consider the computation of a path in which the robot is required to move in a straight line from point A to point B, as in Figure 4–12.

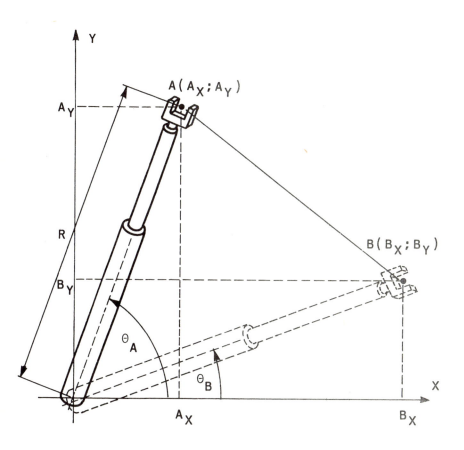

FIGURE 4.12: An example of path computation

We have selected a robot with only two degrees of freedom, to avoid more complicated calculations.

The task of the robot in Figure 4–12 is to move the gripper along a straight-line path. The first of the two degrees of freedom is provided by a revolute joint. Its position in space is defined by the angle between the arm and the axis. The second degree of freedom is provided by a prismatic joint. Its position in space

75

is defined by the distance between the robot base and the center of the gripper.

The center of the gripper, which we will refer to as the tool center point, or TCP, must be moved along the straight line AB.

The solution of the problem of motion along a straight line is provided by dividing the segment AB into many smaller segments, and by calculation of the angle (θ) and the length (R) of the joints at each point along the segment AB.

We will now calculate the values of θ and R required to place the TPC at point A.

Point A is defined in the plane xy according to coordinates Ax;Ay. Therefore, the following relationships exist:

$$A_x = R_A \cos \theta_A$$

$$A_y = R_A \sin \theta_A$$

These formulas permit us to obtain the values of θ and R at point A. Dividing the first formula by the second, we now have:

$$\tan \theta_A = \frac{A_y}{A_x}$$

By squaring these formulas and adding the squares, we get:

$$R_A^2 = A_x^2 + A_y^2$$

Continuing our calculations, we solve for θ and R getting:

$$\theta_A = \arctan \frac{A_y}{A_x} \qquad R_A = \sqrt{A_x^2 + A_y^2}$$

We have now found the values of Θ and **R** required to place the TCP at point A. The values required to place the TCP at point B are:

$$\theta_B = \arctan \frac{B_y}{B_x}$$

$$R_B = \sqrt{B_x{}^2 + B_y{}^2}$$

Motion along the line segment AB will be accomplished by computation of parameters Θ and **R** for each point along the line, and by commands to adjust the revolute and prismatic joints accordingly.

As the number of calculation points along the path increases, the motion will then more precisely resemble a straight line.

Now we can give the example numerical values. Given that the robot has two degrees of freedom, as shown in Figure 4-12, the coordinates of point A are:

$$A = (A_x;A_y) = (2;3)$$

and

$$B = (B_x;B_y) = (5;2)$$

The commands determining joint motion – that is, Θ and **R** – will be calculated by the formulas we derived earlier. To place the TCP at point A:

$$R_A = \sqrt{2^2 + 3^2} = 3.6$$

$$\theta_A = \arctan \frac{3}{2} = 56.3^o$$

Calculating in the same manner, to place the TCP at point B:

$$R_B = 5.3$$

$$\theta_B = 21.8^o$$

In the same way, we can calculate the values of many points along the path.

These points, when put together, will form a straight line.

In order to move the robot along a straight line, the controller must include software capable of calculating points along the path, and of moving the robot joints to pass through all the points. We can achieve motion from point A to point B without such software – that is, by point-to-point control – but we would not be able to control the path between the two points.

As a general rule, the more advanced the robot software, the easier it will be for the user to operate the robot and to write user programs. Here is an example that can be written very simply:

Go to point A.

Approach point B in a straight line.

Stop 50 cm before reaching point B.

And so on.

The controller software will then decipher these commands and translate them into orders to the arm motors.

If the controller software cannot handle more complex input in the programming language, the user will have to move the robot manually from point A to point B, through a large number of intermediate points forming a straight line. This is a difficult and time-consuming process.

Clearly, then, the controller software is a vital part of the robotics system. But it has a further importance. Methods of teaching individual robots how to carry out tasks are determined by the controller software. We will explore this relationship in the next chapter.

Chapter Five

Robot Teaching Methods

PREVIEW

An employee starting a job must first learn the task he is to perform. In this, a robot is no different. Humans can learn assigned tasks from generalized definitions. Robots, on the other hand, must have each task broken down into subunits, and possibly even further, until the analysis reaches the level of individual motions.

In humans, the brain is responsible for learning the procedures to be performed. The brain receives vital information from the senses, such as the location of objects involved in the task. Motions required to carry out the task are dictated to the hand by the brain. The five senses, in turn, provide feedback to ensure that the task is properly completed.

In most of today's industrial robots, the controller is the focus for extremely detailed and explicit commands, compared to a human. Consider, for example, the process of teaching a human and a robot how to build a tower of two blocks.

The human would require a very simple instruction: "Build a tower of two blocks." He would interpret the words, see the location of the blocks, remember how a two-block tower looks, and use his hand almost instinctively to place one block on top of the other.

For a robot, the process must be carefully separated into individual motions, somewhat like the following:

* Move arm until directly above first block.

* Lower arm until block is between gripper jaws.

* Close gripper jaws.

* Raise arm with block held in gripper.

* Move arm until directly above second block.

* Lower arm until block in gripper rests on top of second block.

* Open gripper jaws.

* Move arm away from two-block tower.

In addition to the eight commands listed above, the proper velocity must be defined for each step. More complicated tasks require very precise velocity control, as well as definition of tens, and sometimes hundreds, of different motions. Laying out the learning process may take a lot of time.

Various robot teaching methods exist. In addition, robot manufacturers and researchers are developing additional methods, intended to reduce the amount of time required to learn new tasks and simplify the teaching process.

Ways to teach robots may be divided into two general groups:

*   Methods in which the robot arm moves physically from point to point in the series. These methods are known collectively as teaching.

*   Methods in which the robot does not move physically, in the teaching stage, from point to point in the series. These are known collectively as programming.

Teaching methods include the following types:

*   Teaching in which the robot is moved by means of switches, called teach-in.

*   Teaching in which the robot is moved by hand, called teach-through.

*   Teach-through using a force sensor (defined in the following discussion).

*   Teaching by means of other types of sensors.

Programming methods consist of various types in which the points in the path of motion are defined for the controller without having to move the end effector. They include:

*   Definition of coordinates by the computer.

*   Shifting coordinate systems.

*   Use of vision sensor and pointer.

*   World modeling.

We will now examine in detail the methods by which robots learn their tasks.

TEACHING METHODS

Teach-in

This is the most common teaching method, and it exists in all types of robots. In this method, the human operator moves the robot through a series of points by means of a manual <u>teach pendant</u>, or <u>teach box.</u> The switches on the manual teach pendant are depressed by the operator in a certain order, so that the end effector is moved to each desired point in turn. Whenever the end effector reaches a desired point in the series, the operator presses a <u>record push-button,</u> which causes the controller to record in its memory the position of the joints (angles for revolute joints, lengths of extension for prismatic joints) and the status of the end effector (for example, whether the gripper is open or closed).

The process "Move the robot, record the point" is repeated many times during the teach-in, recurring at each point along the path.

The operator holds the teach box – a small, lightweight object – in his hand. It is connected to the robot controller by a long cable, which allows the operator to move about. In effect, the operator functions as a sensor for the robot, since he is responsible for ensuring that the end effector is properly positioned before pressing the record push-button. Later, when the series is run, the controller calls the points from its memory, and sends movement commands to the joints' motors and to the end effector for each point as required and as it was recorded in the learning stage.

The variety of existing teach boxes is considerable; each robot has a teach box adapted to its particular needs. Teach boxes differ in their levels of sophistication. In the more simple ones, the push-buttons are themselves control motors. In more advanced boxes, the push-buttons program the controller, and may be used to move the robot in straight lines or on various coordinate systems.

Figure 5-1 illustrates a typical manual teach pendant. The push-button functions are of the type which is standard for existing teach boxes.

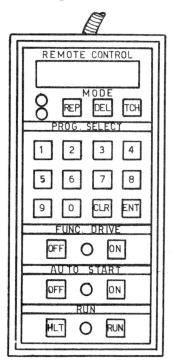

FIGURE 5.1: A typical manual teach pendant

Although the use of manual teach pendants is quite common, the method does have a significant disadvantage: the operator must look away from the robot motion to locate the proper push-button to move the robot. A method which eliminates this problem involves the use of a <u>joystick</u> similar to that used by a pilot to control the attitude of an aircraft.

In this method, the switches controlling robot motion are mounted on the joystick in such a way that moving the stick in a certain direction depresses one or more switches. Moving the joystick to the right, for example, depresses the "right" switch; moving it to the right and forward at the same time depresses both the "right" and the "forward" switches.

Figure 5–2 shows a joystick that moves the robot in only two dimensions. In reality, however, the joystick also controls vertical motion and wrist joint movements.

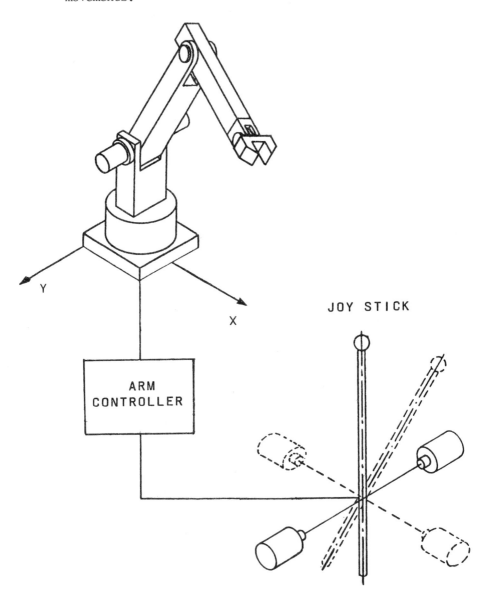

FIGURE 5.2: Teaching a robot using a joystick to control the movement of the arm

Still another teaching method involves the use of an additional robot, called a master robot. This is a lightweight machine that does not include motors or motion-transfer devices. Rather, it has encoders mounted on its joints. Its links are equal in length to those of the slave robot, which will be carrying out the task being learned. The two robots are controlled by a common computer.

During the teaching process, the operator moves the teaching robot through the desired path of motion, which is easily accomplished because this machine is so light in weight. The motions of the master robot are recorded, as in the methods described earlier, so that the controller can issue commands from its memory. Since the geometric structure of the master and slave robots is identical, the path to be traversed by the slave in the future will be exactly the same as the recorded path of the master robot.

Teaching methods involving master and slave robots are used when a high degree of precision is not required, and when it is difficult to break down the overall task into a series of individual points to create point-to-point motion. A typical example of an application of this method is in teaching spray painting. Figure 5–3 shows this task being taught by a master robot.

Teach-Through

In this method, the robot is manually moved along the desired path, and the controller records the joint positions by sampling or taking readings at fixed intervals of time. The robot motors are inoperative; its joints are moved by the human operator.

This method has two main limitations:

* The operator has to overcome the weight of the robot as well as the friction that exists in the arm joints and gears.

*   The controller memory must be very large, in
order to store the information sampled at the
rate of tens of points per second.

The first of these limitations makes the method
inapplicable for medium-size and large robots where
high precision is required. It may be partially
overcome by means of a <u>balance unit</u> which supports the
static weight of the arm.

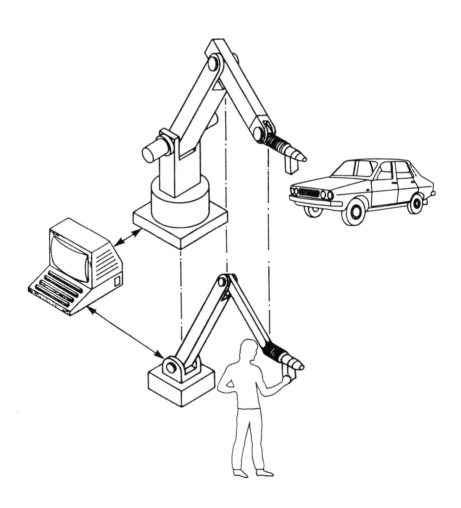

FIGURE 5.3: Teaching a robot the movements required
to spray paint automobile parts using a slave robot

Teach-Through Using a Force Sensor

This method resembles the regular teach–through, but involves the assistance of a force sensor attached to the end effector. When moving the arm, the operator exerts force on the end effector; the sensor translates the exerted force into electrical signals, which activate the robot motors in the desired direction.

The advantage of this method over the regular teach–through is that the operator does not have to expend great effort to move the robot. Therefore, a much higher degree of precision can be achieved. Teach–through assisted by a force sensor is, however, not yet in wide use; its application is limited to experimental models.

Teaching by Means of Other Sensors

In this method, sensors assist the operator in guiding the robot along its learning path by transmitting information directly to the controller. Although it facilitates the learning process, this method is more expensive than the other methods we have discussed because of the sophistication of the equipment.

The sensors discussed here are used during the learning process only, since they are not effective during the performance of the assigned task. For example, in learning a welding procedure, the robot must know the path of its travel along the weld line between the two parts to be joined. This data is supplied by the sensor. However, many sensors are negatively affected by interference generated by the welding process, itself. Therefore, the sensor is used only during the learning stage. However, it is important to note that a sensor capable of functioning during the actual task – that is, one which could resist interference – would make the learning process entirely unnecessary.

In the example given above, the human operator did not move the welding robot along the entire path during the learning process, but only brought the robot to the starting point of the path. The teaching was done automatically as the sensor led the robot. It is therefore obvious that the use of sensors greatly decreases the time and effort required to teach robots their assigned tasks.

With this data in hand, we will now contrast these methods of robot learning with the programming methods of learning tasks which we listed earlier.

PROGRAMMING METHODS

We have stated that, in the programming methods, the robot does not move from point to point in the learning stage. The path of motion is defined for the controller without having to move the end effector. We identified four programming methods: definition of coordinates by the computer; shifting coordinate systems; use of a vision sensor and pointer; and world modelling. We will discuss each of these topics in turn.

Definition of Coordinates by the Computer

Practically speaking, the controller can receive information about a series of points through which the end effector is to pass, and can store them in its memory without actually having to move the end effector along the desired path during the learning process.

This method is used in most robots, and provides a basis for off-line programming control, which we discussed in the last chapter. For example, a task may be defined by a series of numbers, such as 2,3,4 and 5,6,7. When performing the task, the robot will first move the end effector to the point defined by the coordinates $x = 2$; $y = 3$; $z = 4$. It will then move the end effector to the point defined by the coordinates x = 5; y = 6; z = 7. These procedures will be explained by the discussion of coordinate systems in the next topic we will take up.

The ability to define motion by dictating coordinates to the robot controller is the basis for learning methods to be discussed in all the following sections.

Shifting Coordinate Systems

This programming method is applicable for robots performing identical tasks in two or more different work cells, or stations in which the robot is doing some task. The robot is taught to perform the task at one of the work cells, by defining a series of points to the arm controller. The coordinate system is then shifted, enabling the task to be copied at the second work cell.

How do we shift the coordinate system?

Picture a robot drilling holes in sheet metal, in a pattern shown in Figure 5-4.

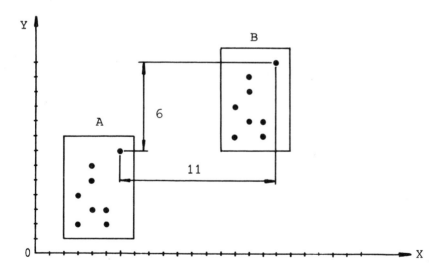

FIGURE 5.4: Shifting coordinate systems

89

In order to save time, two hole-drilling stations are set up. While the robot is drilling at station A, the metal plate at station B, which it has already drilled, is disassembled, and a new one is assembled there. When the robot finishes drilling at station A, it moves on to drill at station B again. Meanwhile the drilling plate at station A is disassembled, and a new one is assembled at that station.

The X-Y coordinates of the holes to be drilled at station B are: (14;8), (14;10), (15;9), (16;9), (16;8), (15;11), (15;12), (17;13).

From Figure 5-4, we see that station B is 11 units distant from station A on the X axis, and 6 units distant from station A on the Y axis. The drilling operation at the two stations is exactly the same.

Given the identical nature of the operations, there is no need to teach the robot to separate series of points for the two stations. All that is necessary is to teach it one series - that at station A, for instance - and then to shift the system of coordinates from station A to station B. The coordinate system shift is defined as follows:

$$X_B = X_A + 11$$

$$Y_B = Y_A + 6$$

During the performance stage, when the robot drills at station A, its path will be exactly as defined by the coordinates stored in the controller memory. When it drills at station B, it adds 11 units to each X coordinate and 6 units to each Y coordinate stored in the memory.

In the above example, the robot has to drill eight holes in each plate. However, to achieve that purpose, it is necessary to teach the robot a large number of points, such as transfer points between holes. In reality, hundreds of points may have to be learned. Therefore, programming by shifting coordinate systems is extremely efficient when identical tasks are to be performed in two different places.

Use of Vision Sensor and Pointer

We have discussed programming by defining coordinates of all the points in the robot path. In order to use this method, we must know the numerical coordinate values of each and every point on the path before we start. However, sometimes these values are unknown to the user, and their measurement involves a lot of effort. The use of a vision sensor helps us overcome this difficulty.

In this method, a narrow beam of light, or a pointer, is used to indicate the points along the path – one point after another. A vision sensor like a camera finds the location of the light spots and inputs them to the controller memory.

A different application of this method involves the use of a monitor screen showing the robot's field of operation. A light pen is used as a pointer; touching the light pen to the desired points on the screen inputs them to the controller memory.

These systems are not yet in common use.

World Modeling

This is an advanced, experimental method, enabling a task to be learned without being broken down into individual motions. It functions by means of overall task definitions, such as "Bring cup" or "Assemble part."

The shapes of all the objects located within the robot's work envelope are stored in the memory. The controller knows where each item is located, its orientation within the space, and how it should be held in order to handle it. Also stored in memory are series of individual movements, each series making up a certain task. This scope of information in the memory enables rapid, easy learning because, as we have said, the task is defined in general, rather than as a set of individual movements. However, we should note that this method of programming requires a rapid computer with a large memory and advanced software. Note that this method is in a development stage at present.

COMPARISON OF TEACHING AND PROGRAMMING METHODS

The advantages of teaching methods over programming methods are as follows:

* The weight of the load transferred by the arm causes the arm to bend to a certain degree. In teaching methods, we can be sure that the arm will reach each point as required, because the same load is attached to the arm during the learning stage as during the performance stage. In programming methods, we cannot know in advance what effect the bending of the arm will have on the location of the end effector at any given point.

* In programming methods, the coordinates of each point must be identified. Inaccuracies in coordinate definitions can lead to a lack of precision in performance.

The advantages of programming methods over teaching methods are as follows:

* Programming methods save time and effort. Moreover, there is no need to take the robot off production to teach it a new task. Programming the new path is done off-line, while the robot's regular work activity proceeds.

* A task learned by defining the coordinates of the points along the path can be communicated

to other robots, even to some that are different in structure.

As we have indicated at several times in our discussion, robots have a long way to go to reach the precision we would like to see. For example, means must be developed to keep the arm position independent of the load. We must devise ways also of preventing the robot from colliding with objects in its path. These improvements will occur. And along with them, we can anticipate that programming methods will become the main techniques for instructing robots.

Chapter Six

The End Effector

INTRODUCTION

> The term "end effector" is a generic word
> for all the systems mounted at the end of the
> robot - that is, at the end of the link
> farthest from the robot base - whose task is
> to grip objects, or tools, and/or transfer
> them from place to place.

Examples of end effectors include grippers, welding
guns, and paint sprayers. The operation of the end
effector is the final goal of robot operation. All the
systems mentioned in our previous discussions - drive

units, controls, and so on – are designed to enable the operation of the end effector. Failure of the end effector will cause the entire task to fail. Thus, it is crucial that the end effector be properly designed and adapted to conditions likely to develop in the environment in which it is to carry out work.

Two main types of end effectors exist: grippers and tools. In this chapter, we will describe grippers. Tools will be discussed in a later chapter on applications.

The gripper is comparable to the human hand. The structure of the hand is truly amazing; it includes 22 degrees of freedom, various drive units in the form of muscles, and many sensing elements. It is so complex and so highly developed that no gripper capable of duplicating its functions will be produced in the near future. Given this fact, it is obvious that grippers are suitable for a limited range of operations.

However, the wide range of demands has led to the development of grippers that can handle objects of many different sizes, shapes, and materials. They fall into several type classes, most of which are adapted for one specific operation.

TWO-FINGER GRIPPER

The most common type is the two-finger gripper. A number of varieties have been created, which differ from each other in size and/or in finger motion – that is, whether they move in a parallel or rotary mannner. Examples of two-finger grippers are shown in Figure 6–1.

The basic disadvantage of the two-finger gripper is the limitation it places on the size of objects it can handle; they must not be larger than the span between the gripper fingers when fully open.

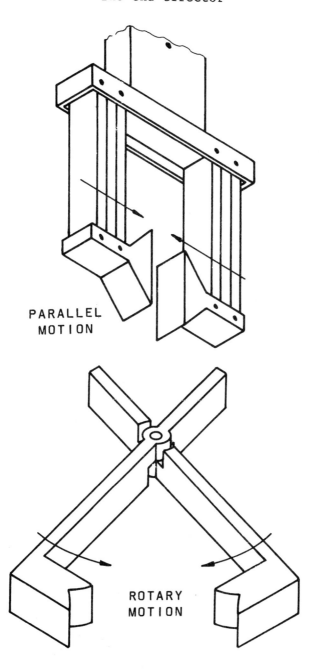

PARALLEL
MOTION

ROTARY
MOTION

FIGURE 6.1: Examples of two-finger grippers

THREE-FINGER GRIPPER

These are basically similar to the two-finger grippers, but allow a more secure grasp of objects which are of a circular, triangular, or irregular shape.

Figure 6–2 shows a <u>three-finger gripper.</u> The fingers shown are jointed, meaning that they consist of several links. However, some three-fingered grippers have fingers like those shown in Figure 6–1.

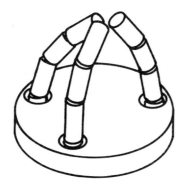

FIGURE 6.2: A three-finger gripper

CYLINDRICAL OBJECT GRIPPERS

This gripper consists of two fingers, each marked with several semicircular indentations. When closed, the gripper can hold cylindrical objects of several different diameters.

The <u>cylindrical object gripper</u> has two disadvantages:

* Its weight, which must be borne by the robot throughout its operation.

98

*   The limitation of movement caused by the length of the gripper.

Figure 6–3 shows a gripper of this type, capable of holding cylindrical object of three diameters.

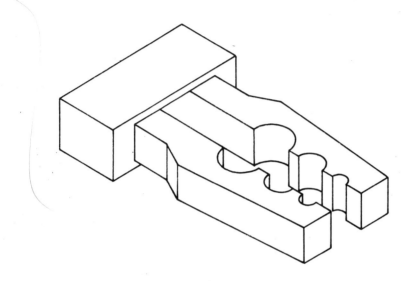

FIGURE 6.3: A gripper for picking up and holding cylindrical objects

FRAGILE OBJECT GRIPPERS

In order to grip an object without dropping it, the gripper fingers must exert a certain degree of force. That force, concentrated at a single point on the object, is liable to damage delicate parts. Numerous fragile object grippers have been developed, one of which is shown in Figure 6–4. It includes two flexible fingers, which bend inward to grasp a fragile object and which are controlled by compressed air.

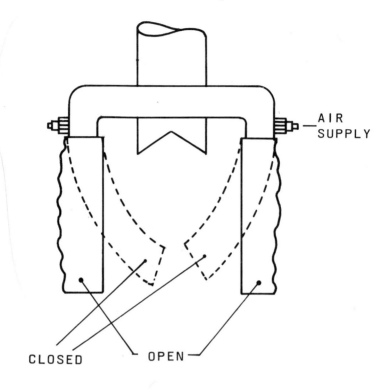

CLOSED   OPEN

FIGURE 6.4: A gripper for handling fragile objects

JOINTED GRIPPERS

These grippers are designed to hold objects of varying
sizes and of uneven shapes. The links are moved by
pairs of cables. One cable of each pair flexes the
joint, and the other extends it. To grip an object,
the fingers wrap themselves around the object and hold
it firmly. The large number of links enables the
gripper to grasp objects of irregular size and shape.
Figure 6-5 shows a typical jointed gripper.

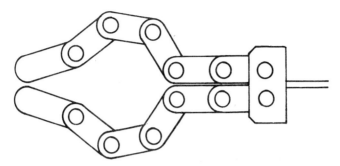

FIGURE 6.5: A jointed gripper having a whole chain of finger segments

VACUUM AND ELECTROMAGNETIC GRIPPERS

Vacuum grippers are designed to attach to flat surfaces by creating a vacuum and to remain attached to those surfaces as long as the vacuum exists.

Electromagnetic grippers are designed to attach to ferromagnetic objects by creating an electromagnetic field.

Both of these gripper types are very efficient, since they can grasp objects of various sizes, and do not require great precision in locating a gripper point.

Vacuum grippers are used to handle objects with planar surfaces, such as metal plates and cardboard boxes. Generally, each vacuum gripper includes a number of suction cups, each of which is connected to an air pressure pump. To reduce the risk of malfunction due to loss of vacuum, it is a common practice to use more than one pressure suction cup.

Figure 6-6 shows a vacuum gripper with four suction cups in the act of lifting a flat object.

Electromagnetic grippers are used to hold objects made of materials that can be magnetized, such as iron and nickel. These objects must have a specific spot at which the gripper can attach itself.

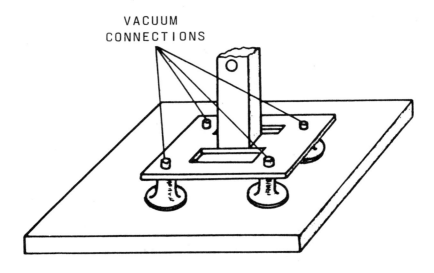

FIGURE 6.6: A vacuum gripper with four suction cups

AUTOMATIC GRIPPER CHANGERS

Many types of work involving robots call for the ability to grasp objects of varying shapes and sizes. In addition, the use of a tool, such as a screwdriver, may alternate with the use of a gripper. To date, no gripper capable of holding all types of objects has been invented. Therefore, it has been necessary to create a unit called an automatic gripper changer. This is actually an adapter which allows a gripper to be quickly attached to the end of the robot arm and just as quickly removed, as required by the operation. An obvious restriction is that all such adapters must attach to the robot arm in the same way and must connect in an identical manner to their drive units, whether electrical, hydraulic, or pneumatic.

There are a number of disadvantages in the use of an automatic gripper changer:

The end effector

* Weight is added to the end of the robot arm.

* Technological complications are a potential source of malfunction.

* Robot cost is increased.

* Time is wasted in changing grippers.

From these observations, it is evident that the development and production of grippers is one of the important stages in the design of robots for particular tasks.

Generally, users do not purchase grippers "off the shelf." Each task requires a gripper designed for that operation. Therefore, manufacturers normally sell robots without end effectors; the grippers or tools are chosen and adapted by the engineering team that installs the robot at the plant. This is a critical stage of installation, requiring a high level of knowledge and skill.

We have now reached the end of a series of chapters on the components of the robot arm. In the next chapter, we will pull together the various phases of robot structure and discuss their integration in actual work situations.

Chapter Seven

Integration and Operation of Robot Subsystems

Throughout the last few chapters, we have presented the various subsystems making up the industrial robot. We will now consider how the operations of these subsystems are integrated into operation of the entire robot. For this purpose, we will assign a task to the robot, and follow the operation of the various subsystems during the performance of that task.

A ROBOT TASK

The task is to build a tower from three blocks of different sizes, as shown in Figure 7-1. The location of the blocks before assembly of the tower is specified, along with the instruction that the tower is to be built on block 1 in its present location.

The commands required to place one block on top of another are:

105

* Go to point 1.

* Go to point 2.

* Close gripper (to pick up block).

* Go to point 3.

* Go to point 4.

* Go to point 5.

* Open gripper (to put block down).

* Go to point 6.

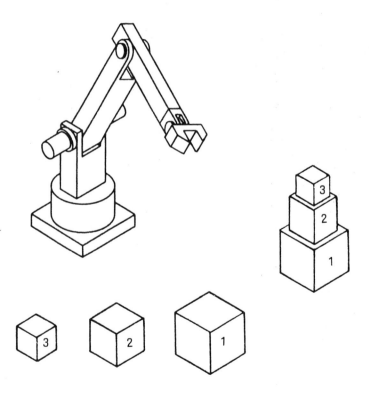

FIGURE 7.1: The task involves placing block 2 on to block 1 and block 3 on to block 2

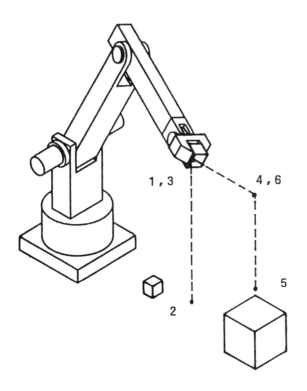

FIGURE 7.2: The path of the robot arm required to
place block 2 on to block 1

Points 2 and 5 are those at which the robot performs an
operation on the block. Therefore, it is very important
that the gripper be precisely located at these points.

Points 1, 3, 4 and 6 are included in the path to create
the vertical motions required to bring the gripper down
to the blocks and back up. These vertical motions are
important, because they prevent the gripper from
damaging the blocks while in motion. However, no
gripper operation is carried out at these points, so
the precise location of the gripper at each of these
points is not important. They are referred to as dummy
or via points.

Placing Block 2 on Block 1

We will teach the robot to go to the points shown in
Figure 7-2 by pressing switches on a manual teach
pendant. When the robot reaches the desired point, we
depress a recording switch; the controller then
records in its memory the position of each joint (angle
and link lengths) of the robot arm, as well as the
status of the gripper (open or closed). The velocity
is also defined for each point.

At the end of the teaching stage, the controller will
have stored in its memory eight rows of data, one for
each motion command. It will list the position of each
joint, the position of the gripper, and the velocity.
In the illustration in Figure 7-2, the robot moves only
three joints - the base joint, the shoulder joint, and
the elbow joint. The series of commands stored in the
controller memory will probably look like this.

Command	Base Joint Angle	Shoulder Joint Angle	Elbow Joint Angle	Gripper Position	Velocity
1	10	70	110	0	15
2	10	60	95	0	5
3	10	60	95	1	
4	10	70	110	1	5
5	0	40	150	1	15
6	0	35	145	1	5
7	0	35	145	0	
8	0	40	150	0	5

Note the following facts about the chart we have just
displayed:

* The numbers are not exact, and are meant to
illustrate a principle only.

* The gripper positions are: 0 = open; 1 = closed.

* The velocities vary for a reason. High velocities reduce path accuracy. Therefore, it is common practice to move rapidly between two dummy (via) points and to reduce speed when moving to a point located near an object or when performing a motion involving work on an object. Path accuracy is not important between dummy (via) points, because there is no danger of damage to the environment.

Placing Block 3 on Block 2

The path required to place block 3 on block 2 will be taught by defining the Cartesian coordinates of the points along the path for the robot arm controller, as shown in Figure 7-3.

In this instance, we do not move the robot arm from point to point along the path. Rather, we define the points by off-line programming. Therefore, we must define the coordinates of the points along the path with great precision – particularly the point at which the robot grips block 3 and the point at which it puts block 3 down on top of block 2. The series of commands stored in the controller memory will look like this:

Command	X	Y	Z	Gripper position	Velocity
9	9	0	16	0	15
10	9	0	1	0	5
11	9	0	1	1	
12	9	0	16	1	5
13	11	5	16	1	15
14	11	5	7	1	5
15	11	5	7	0	
16	11	5	16	0	5

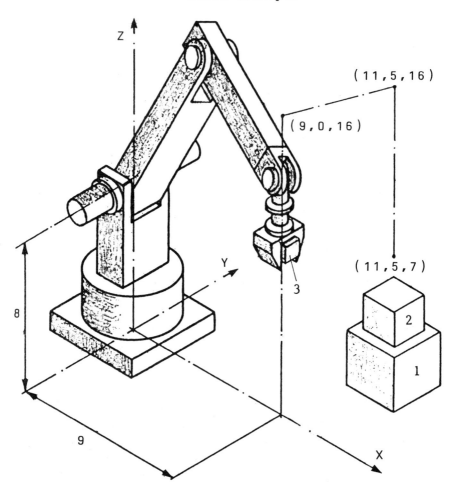

FIGURE 7.3: The path of the robot arm required to place block 3 on to block 2; the definition of coordinates in space

This stage completes the learning process of the task. The series is stored in the controller memory, and will be called up whenever we want to run it.

Implementation of the Series

The next step is implementation of the series, or running it. The human operator instructs the controller to implement the series, which he designates by a name he has given it. Once the controller has received the instruction to run the series, it addresses its memory and calls up the first row of the program – that is, the position of the joints (10, 70, 110); the position of the gripper (0); and the velocity (15). By means of the encoders the controller reads the position of the robot joints before starting any motion. It then calculates the difference between the position dictated in command 1 and the current position. Following this calculation, it moves the arm to the position dictated in command 1. This process is repeated from point to point throughout the series.

Consider, for example, the process involved in executing command 2. The base joint angle does not change, the shoulder joint angle decreases by 10 degrees, and the elbow joint angle decreases by 15. The gripper position remains unchanged (open), and the velocity is defined as 5 units. The robot controller recognizes that it has to alter the position of two robot joints in order to carry out the command.

Whenever the controller computer recognizes that the desired values of the robot joints do not equal their current values, it issues motion commands to the joints, in digital form. A digital-to-analog converter translates the digital data into the required analog signal. A driver amplifier then steps up the signal to activate the drive units. The drive units move the joints directly (in the case of direct drive) or by means of gears, shafts, belts and so on (in the case of indirect drive). The combined motion of the joints creates motion of the robot arm, in order to move the end effector to the required points.

When executing commands 9 through 16, the arm controller must first translate the Cartesian coordinates listed in its program into the appropriate joint variables. Once translation is completed, implementation continues as for commands 1 through 8. Translation of the point location from Cartesian coordinates to joint angles must be carried out by the controller for each motion to a new point. In paths consisting of straight-line motion between two points, translation calculations are carried out many times during the motion, for many points on the straight line. The more points for which calculations are carried out, the more closely the arm motion will resemble a straight line.

An important part of the robot control process is regulation of the arm velocity. When the end effector approaches the point at which the arm must stop, a controlled slowing down must take place to mark the transfer from position velocity (that is, motion) to zero velocity (that is, rest). A sudden stop, without prior slowing down, will jolt the end effector at the stopping point, reducing the precision of the robot's operation. In today's industrial robots, velocity control is automatic; when the end effector comes within a certain distance from the point where it is to stop, as defined in the software, a gradual slowing takes place, ending in a total stop at the desired point.

Now, having covered the principles of operation of the various robot subsystems in the last several chapters, and having discussed their integration in this chapter, we can look at robot performance in a larger sphere – that is, as they are used in overall industrial processes. Robots are not isolated factors in industry. Rather, they are integrated components of work cells, in which they may perform production operations or may feed other machines. The functions of robots in industry, and the types of work they perform, are covered in the next chapter.

Chapter Eight

Industrial Applications of Robots

INTRODUCTION

As we indicated early in our discussion, robots
are being applied in industry to an increasing
number of jobs which are, on the one hand,
hazardous to human life, and, on the other hand,
merely boring or physically difficult. In this
chapter, we will explore the most common of these
applications.

In each instance, we will discuss the process of
selecting a robot and adapting it to the work
station. We will then present an example of the
application and examine the characteristics of the
robot which are necessary to carry out that particular
job.

INDUSTRIAL APPLICATIONS

The applications to be described fall into the following categories:

* Press loading and unloading

* Die casting

* Machine tool loading and unloading

* Spot welding

* Arc welding

* Spray painting

* Assembly

* Finishing

Press Loading and Unloading

Pressing is an operation used in forming and reshaping parts. The workpiece is placed in a press which exerts extreme pressure on it or cuts portions away from it in order to give it a new shape. The transfer of pressure from the press to the workpiece is accomplished by a special form called a die; the unmachined part is placed within the die and assumes the shape of the die. Normally, the part goes through a number of pressing operations to receive its final form.

Pressing operations are carried out in many types of plants. Presses are used in the machine-making industry to form the exterior metal portions of machines. Many screws are manufactured by pressing. And even such parts as the blades of jet aircraft are formed by a series of pressing operations.

Figure 8-1 shows a typical work cell for a pressing operation, as well as the robot that performs the work.

114

The robot takes an unmachined part from the part feeder, places it in the press, and transfers it from the press to a conveyor belt or to a <u>pallet</u> for finished parts at the end of the pressing operation.

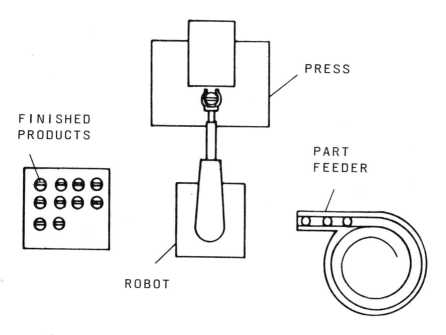

FIGURE 8.1: A typical layout for robotic pressing operations

The motions required in this application are simple, and the exact path between the machines making up the work cell is not important. Therefore, the robots used in pressing operations are first-generation, "pick and place" robots.

Actually, a process in which one robot picks up a machined part from the feeder, loads it on the press, removes it when it has been formed and places it on the pallet, then returns to the feeder to pick up another unmachined part would be too complicated and time-consuming. Instead, an arm with two grippers is used for each of the two phases of the operation – one for loading the press, and the other for unloading it onto the pallet.

Users of robots for press loading and unloading report a number of advantages from integrating robots into the pressing process:

* Reduction in the number of humans required to operate the system.

* Increase in production due to the ability to work two and three shifts.

* Significant reduction in injuries, both while operating the press and while holding parts.

* Increase in the number of parts manufactured per unit of time.

* Improvement of working conditions for humans, whose jobs have changed from performing monotonous and hazardous tasks to supervising production machines.

There are still numerous problems in the integration of robots into the pressing process. Basically, they concern cases where the system fails − for example, when the raw material runs out, or when a machined part gets caught in the die and the robot is unable to pull it out. One attempt to overcome these problems involves the use of a simple sensor in the robot arm. The sensor reports to the arm controller that a problem situation exists; the controller then shuts down the robot and calls a supervisor.

Die Casting

This operation is performed by injecting a raw material at its melting temperature into a special form, or die. Inside the die, the material cools and solidifies, thus conforming to the interior shape of the die. The die can be opened to take out the cast part after it has hardened. Generally, the process is used to manufacture parts of raw materials with low melting temperatures − aluminum, lead, or plastic − so that the die can be made of a readily available material that has a higher melting point.

Casting, like pressing, was one of the first operations to integrate robots – in the early 1960's. This was due to the nature of the operation, which involves simple, often-repeated motions under difficult environmental conditions (heat, noise, and risk).

After injection, the casting process includes the following stages, as shown in Figure 8–2:

* Removal of the hardened part from the die casting machine.

* Transfer of the part to a quench tank.

* Removal of the cooled part from the quench tank and transfer to a press for final shaping and trimming.

* Transfer of the completed part to a pallet or conveyor.

Some casting processes include insertion of parts made of other materials into the cast part, to increase the mechanical strength of the finished part. In this case, the robot's work includes transfer of the additional part from a feeder and precision positioning of the additional part in the die before injection of the melted raw material.

Figure 8–2 shows a casting work cell, with a robot tending a casting machine and a press. The operation shown includes the insertion of an additional part into the die before the actual casting occurs; therefore, the work cell also includes an additional parts feeder.

To prevent malfunctions, the robot controller is connected to the die casting machine and to the press. It supervises the timing of the various operations, so that the machines do not lag behind the robot, or vice versa. For example, the press should be activated only after the machined part has been placed inside it and the robot arm has been moved away from the press.

Likewise, the robot arm should not approach the die casting machine until the die has opened.

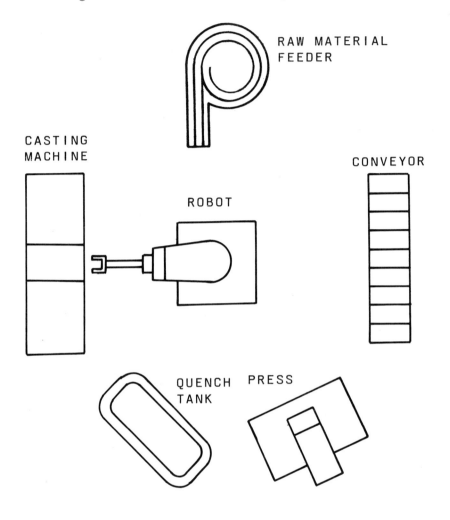

FIGURE 8.2: A layout of a typical die casting/pressing work cell

The advantages resulting from the use of robots in die casting are similar to those achieved in pressing: reduction in personnel, increased production from the ability to work shifts around the clock, reduction of injuries, and increase of output.

In die casting, as in pressing, a human operator must work directly with the system, to supervise the machines and to solve problems.

Machine Tool Loading and Unloading

Machine tools are those that machine parts – lathes, milling machines, grinding machines, and so on. Prior to the introduction of robots, human operators usually performed the following operations:

* Loading and unloading the material.

* Setting the machine at the conclusion of each stage in the machining process.

* Inspecting the dimensions of the machined parts.

* Changing the tools in the machine.

* Troubleshooting.

With the introduction of CNC machines, the need for skilled machining workers was reduced. In this type of production setup, the human operator does little more than load and unload the machine.

For some time, relatively few plants adopted robots for this loading and unloading, because it was thought to be too costly. A robot working with a single machine, and doing approximately the work of a human operator, would be "unemployed" most of the time, while it waited for the end of the machining process.

However, the integration of a robot into a work cell in which it serves a number of machine tools has proven to be cost–effective. Therefore, a significant increase has occurred in the last few years in the number of robots used to load and unload machine tools.

Figure 8–3 shows a robot servicing two lathes and a milling machine. The arrangement of machines and finished part pallets makes up a flexible unit which inputs raw material, machines it stage by stage, and outputs a finished product. The total unit – machines, pallets, and robot – is called a <u>production cell</u>.

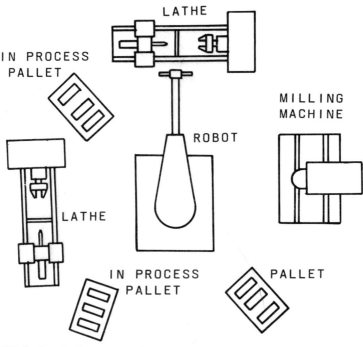

FIGURE 8.3: A layout of a production cell – a robot servicing two lathes and a milling machine

In order to use one robot in loading and unloading a number of machine tools, the operations of both robot and machines must be precisely timed. To achieve this objective, the work station must be carefully designed to permit the placement of all relevant machines and parts feeders within the robot work envelope, and all motions must be planned to avoid collisions between the robot and the surrounding equipment.

Mobile robots are capable of reaching and servicing large numbers of machines. To facilitate such an operation, robots may be installed, for example, on <u>gantry</u>-type overhead tracks, along which they move from machine to machine. Figure 8–4 shows a robot moving along a gantry track and servicing several machine tools.

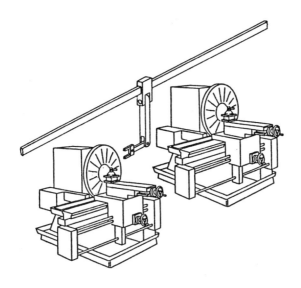

FIGURE 8.4: A robot servicing two machine tools by moving along a gantry track

Once more, it is important to note that robots cannot replace human operators entirely in loading and unloading, as in pressing and casting. Some tasks, such as inspecting work and repairing damaged products, must be performed by humans.

Spot Welding

Spot welding was one of the first tasks for which robots were integrated into production systems. Welding is difficult, monotonous work, requiring a high degree of precision. Robots are ideally suited to this task. Their motions are precise, and their wrist joints enable them to reach every location on the workpieces without damage to the parts or to the robot arms. In addition, the flexibility of robot work stations permits the storage of various welding programs for different production jobs, as well as for rapid modification of programs.

The process of spot welding is based on a high current flowing between two electrodes and through two pieces of metal which are to be joined. As the current flows, extreme heat is generated at the point of contact. Due to the electrical resistance of the metal parts, this heat causes the metal to melt at the contact points. The pressure of the electrodes is maintained for a short time after the current ceases to flow, in order to hold the metal parts together while the welded spot cools and solidifies. The electrodes themselves are prevented from melting during the current flow by a cooling fluid that flows through them. Figure 8-5 shows a welding machine and demonstrates the process of spot welding two metal parts.

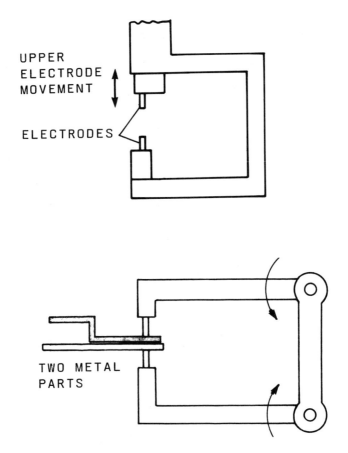

FIGURE 8.5: Spot welding

Applications like the one depicted in Figure 8–6 can be seen in most automotive plants today. A car is transported on a conveyor to a work cell whose assigned task is spot welding. Up to 12 robots are positioned on both sides of the conveyor, and are set to work on the car as soon as it is brought into position by the conveyor. Knocking noises are heard. Sparks fly. And within minutes, the car is transported out of the work cell with its body welded at hundreds of different spots.

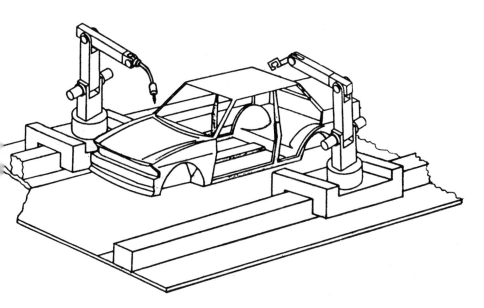

FIGURE 8.6: Two robots spot welding a car body

Today, automotive plants use more robots than any other type of industry, not only for welding, but for spray painting and assembly operations. About 30% of the world's robots are currently used in the automotive industry.

The spot welding gun – the proper name of the end effector used for this job – is relatively heavy, and the required work envelope is relatively large.

Therefore, spot welding is done by large robots capable of carrying a <u>payload</u> of 35 pounds and up.

Another characteristic required of spot welding robots is the ability to perform complicated motions. This is because the robot must be able to follow contours in the workpieces and to reach inaccessible points without damaging the parts being welded. In addition, the process requires that the electrodes be perpendicular to the workpieces – meaning that the spot welding gun must be repositioned at each welding point to maintain this posture.

Most spot welding applications use robots with six degrees of freedom – three for positioning and three for orientation, or posture at the workpiece.

Although the motions required in spot welding are complicated, the only point where great precision is necessary is that of the actual weld; the path travelled by the robot between welding points is unimportant. It is therefore possible to use point-to-point control robots for spot welding, rather than the more sophisticated continuous path control robots. To prevent collisions between the robot and the pieces being welded during motion between welding points, the robot is taught a considerable number of via (dummy) points through which it must pass on its way from each welding point to the next.

Teaching spot welding tasks is a complicated process. The robot must be manually transported through each of the hundreds of welding points, and must be positioned with a precision of +/- 1 millimeter. Since the electrodes must be perpendicular to the workpieces, it is extremely difficult to achieve this precision. In automotive assembly, for instance, the fact that a number of car models are produced on the same production line requires the teaching of a new series of welding points for each model. Therefore, the teaching process involved in robot-performed spot welding of car bodies is long and wearying.

The teaching process may be simplified by the use of smart software, which can, for instance, enable a change in orientation of the welding gun without changing its location in space. This permits the welding gun at the end of the robot arm to be brought to the desired location and then turned to the desired orientation without having to correct for location. The robot control mechanism, in this case, corrects the robot joint positions automatically, so that the position of the welding gun is constant.

Operations included in the spot welding process are:

* Rapid motion of the robot arm, with welding gun attached, to the vicinity of the point to be welded.

* Bringing the welding gun electrodes to both sides of the part to be welded and positioning them exactly opposite the welding point.

* Attaching the electrodes to the point to be welded.

* Running an electric current through the electrodes and the welded material.

* Waiting.

* Opening the electrodes.

* Moving the robot arm to the vicinity of a new welding point.

The main advantages involved in using robots in spot welding are:

* Better weld quality.

* Precise positioning of the welds, ensuring strength.

* Saving of manpower and time (about 1.5 seconds per weld).

The main disadvantages are the failures in the process that occur because of physical deterioration of the electrodes and the tedious teaching process.

Arc Welding

This is a process used to join two metal parts along a continuous contact area, rather than at individual contact points, as in spot welding. In arc welding, the two metal parts are heated along the contact area until the metal melts; as it cools, the molten metal solidifies, joining the two parts.

To create an electric current, two electrodes with potential differences between them are required. The spot welding gun includes two electrodes, between which current flows. By contrast, the arc welding gun has only one electrode, with the metal object to be welded serving as the second electrode. The potential difference between the two electrodes, necessary to the generation of the electric current, is supplied by the welding equipment system.

The metal objects are heated by an electric current, which flows through the electrodes in the welding gun and through an air gap to the object being welded. When using a robot in arc welding, the welding gun is attached to the end of the robot arm, and the electrode is fed through the gun, in the form of a metal wire. The welding gun also disperses a special gas to prevent oxidation of the heated area, which would adversely affect the weld quality.

The arc-welding process requires the use of high-quality robots with sophisticated software, capable of performing the following operations:

* Rapid motion to the vicinity of the contact area to be welded.

* Transmission of signals to cause dispersal of the gas, application of voltage to the electrode, and feeding of the electrode wire.

126

* Precise motion along the welding path while maintaining a constant gap of air.

* Preserving a constant orientation of the electrode relative to the surface to be welded.

* Keeping the welding gun moving at a constant velocity.

* Ability to perform weaving motions, in order to achieve a good joining between the two metal bodies and improve the welding quality.

To meet the above requirements, arc welding requires robots with the following characteristics:

* Five to six degrees of freedom.

* Continuous path control, for exact motion along the welding path and regulation of velocity.

* High repeatability. This is a quality related to the radius of the sphere around any previously learned point within which the robot will locate its end effector when performing the assigned task. For example, the smaller the circle, the higher the repeatability. Repeatability is used to measure the precision with which the robot carries out previously learned tasks.

Manual arc welding demands highly skilled human operators, capable of controlling the velocity of the welding gun, the size of the air gap between the welding gun and the metal workpiece, and the precision of the welding path, all at the same time. Therefore, it is to be expected that the quality of arc welding performed by a human, expert though he may be, will be lower than that performed by a robot, because of the greater precision of which it is capable.

By its very nature, arc welding is an unpleasant job for humans, due to the high temperatures and to the fact that human welders must wear masks to protect their eyes from the strong light generated by the welding process. Special clothing must be worn to protect the worker's body from radiation and sparks. But, in spite of these conditions and in spite of the superiority of robots over humans in process precision, arc welding is still done by humans in many places. This is mainly because of the complexity and sophistication required of robots assigned to such tasks.

Problems resulting from the integration of robots in arc welding include:

* Using the teach–in method in arc welding is a laborious process, especially in the case of curved paths.

* Teaching arc welding by teach–through methods is difficult for the individual who is moving the arm along the path manually. Also, his motions cannot be exactly identical to those to be carried out by the robot during its performance runs.

* When the metal bodies are heated, distortion occurs, causing the join line to displace slightly during the welding process; it is no longer precisely identical to the learned path. This situation is not obvious in short welds, but can be quite significant in longer ones as well as in places where heat is not rapidly dissipated from the welding area.

The above problems may be solved by the use of a sensor to identify the join line between the parts. The use of a sensor eliminates the need for a learning stage; the sensor guides the welding gun along the join line during the welding process, in spite of the light and heat generated by the welding. We will discuss arc welding sensors and their problems in Chapter 9 when we discuss sensors.

Figure 8-7 shows an arc welding work station. It illustrates the components necessary to the welding process, as well as jigs and a rotary index table used in the precise positioning of the parts to be welded. The use of two rotary index tables allows the robot to work continuously, performing welding at one table while loading and unloading is occurring on the other.

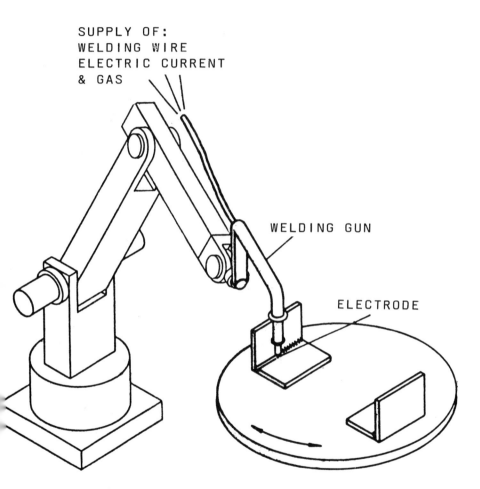

SUPPLY OF:
WELDING WIRE
ELECTRIC CURRENT
& GAS

WELDING GUN

ELECTRODE

FIGURE 8.7: A rotary index table may be used to obtain accurate positioning during arc welding

The feeding unit regulates the feed rate of the welding wire, which is adjusted to the current and voltage measured at the voltage source of the welding gun.

This feed rate regulates the width of the air gap between the wire and the part to be welded.

In summary, then, the main advantages of using robots in arc welding are:

* Improvement of weld quality by eliminating the human factor and its negative influence on product quality.

* Reduction of work time, especially when welding along a short path.

* Reduction of costs, due to the limited use of highly skilled labor.

* The ability to work continuously, whereas human operators must rest occasionally because of the difficult working conditions.

* Improvement of working conditions for humans, who no longer work in high temperatures nor wear protective masks and clothing.

Spray Painting

The use of robots in spray painting involves the attachment of a spray gun as an end effector, which is moved along a previously learned path. Characteristics typical of painting robots include:

* Continuous path control.

* Rapid motions.

* Low repeatability (0.1 to 0.2 inches)

The task assigned to a painting robot is to repeat motions taught by a human painter, while at the same time operating a spray gun along the path. The spray gun is attached to an air supply.

Industrial applications of robots

Spray painting is taught by the teach-through and by the teaching or master robot methods.

In that connection, the flexibility of robots – that is, their ability to work on various types of parts – becomes evident in spray painting applications. Painting robots can store an appropriate program for each part type, and can call the correct program from memory in response to the part to be worked on.

Most robots used in such applications are <u>blind</u>. Therefore, the part to be worked on must be positioned at a point familiar to the robot – that is, a point located at a given distance and direction from the robot base. Work may be performed on objects at rest or in motion. In the case of stationary objects, the robot begins operation only after receiving a signal confirming that the part is properly positioned. When working on mobile objects, the robot receives signals from the conveyor used to transport the part. These signals, which continue as long as the part is in motion, update the robot as to the distance and direction of the part from the base.

A convenient arrangement for spray painting moving parts involves suspending the parts to be painted, as shown in Figure 8-8. This procedure allows the painting of the entire surface area, except for the points of suspension. Also, excess paint runs off into a catch basin without dripping onto the conveyor or interfering with its operation. Suspended parts must be hung in such a way that they cannot swing or rotate while in motion so that their location does not deviate.

The main advantage in the use of robots in spray painting include:

* Rapid return on investment. Since specifications for these robots do not call for precision or large physical size, they are not expensive. In addition, they do their work rapidly without tiring, and save considerably on paint costs by spreading the paint uniformly.

* Improvement of working conditions for humans. Spray painting is both difficult and dangerous. Since the air in the work area is mixed with paint fumes, human painters have to breathe through special filters while on duty.

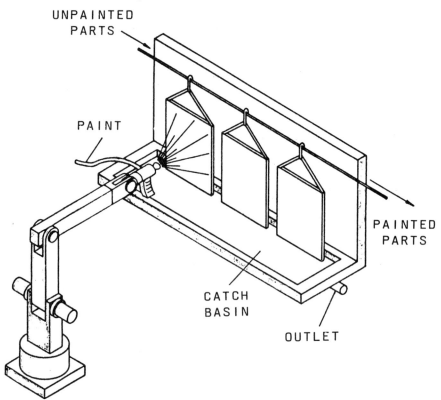

FIGURE 8.8: An arrangement which allows moving parts to be spray painted

The principal problems involved in the use of robots in this application include:

* The need to protect painting robots from fumes and dirt.

* The need to isolate any electric sparks produced within the robot during its operation, since the painting environment is extremely flammable.

* The need for coordination between robot motions and the location of the parts to be painted.

* The fact that some spray painting operations involve surfaces that are not easily within reach. In some applications, the robot must have more than six degrees of freedom to overcome obstacles and reach remote surfaces.

It is interesting to note that a similar application is gluing. The process involves spreading a layer of glue along defined lines in a manner resembling spray painting. It is gaining popularity as a result of an improvement in the quality of glue and in gluing methods. Today, many products - even aircraft - include glued-on parts.

Assembly

Great emphasis is presently being placed on the development of robots capable of performing assembly operations. The main reason is that nearly 40% of manpower costs in production come from assembly.

The automation of assembly operations is not simple. Solutions must be found for a number of requirements, among which are:

* High degrees of precision and repeatability in positioning the end effector.

* Motion in straight lines while maintaining fixed orientation of the end effector.

* Automatic end effector changing, or a versatile gripper.

* Rapid motion of the robot arm.

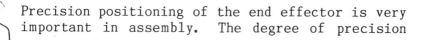

Precision positioning of the end effector is very important in assembly. The degree of precision required is measured in hundredths of ten-thousandths of an inch. Robots used in such applications must be capable of precision greater than the allowed tolerance of the assembly. Precision must also be maintained in the orientation of the end effector, to ensure that the assembled part is held at the correct angle.

TRANSLATION ERROR   ANGULAR ERROR

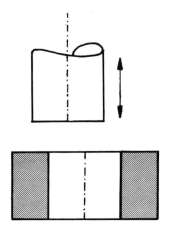
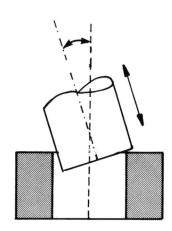

FIGURE 8.9: Misalignment of parts during an assembly operation

Figure 8-9 shows a lack of precision in the location of the assembled part (a translation error) and in the orientation of the part (angular error). Precision of position and orientation is not easy. Several methods of achieving these problems will be discussed later in this chapter.

Most assembly operations involve a downward linear motion, while maintaining precise and constant orientation of the assembled part. Therefore, assembly robots must always include software enabling them to execute linear motion.

In many assembly operations, several parts of different sizes and shapes are assembled onto one central part. If this task is to be carried out by a single robot, it must include an end effector capable of holding dissimilar objects, or be able to change grippers in the middle of an operation, with the aid of an automatic tool changer. The use of an automatic tool changer is preferable when the robot is working on a part rather than merely handling it. An example is driving screws during assembly. In such a case, one of the tools attached to the automatic tool changer would have to be an automatic screwdriver.

The velocity of robots used in assembly is greater than that required in almost any other industrial application. A velocity in the order of 1.5 to 3.5 feet per second – the greatest velocity of which most medium-sized robots are capable – permits assembly at a rate comparable to that of manual assembly operations. Faster robots can reduce assembly time, especially if the robot path includes long segments of motion between points. However, it should be remembered that arm motions during assembly – that is, while actually fitting one part onto another – are not carried out at maximum velocity. A lower velocity not only prevents damage to parts, but increases precision.

The requirements we have just discussed add to the complexity of assembly robots, and therefore to their cost. On the other hand, since most components involved in assembly weigh less than 2.5 pounds, their use is simplified and their cost is reduced. At the same time, greater precision and rapidity can be achieved.

Assembly methods may be divided into two broad categories:

* Assembly of parts in a vertical direction. An example is the assembly of electronic components onto printed circuits, which involves only downward motion.

* Assembly of parts in different directions. An example is the assembly of automobile engines.

Applications requiring assembly along a vertical axis can be accomplished by Cartesian, cylindrical, or horizontal articulated robots with four to five degrees of freedom. These robots are easily capable of motion along a vertical axis while maintaining a constant orientation and a high degree of precision, since the vertical motion is linear and is accomplished by a single motor.

Assembly of parts in different directions requires robots with six degrees of freedom, since they must maintain linear motion along various axes while remaining in constant orientation to the end effector.

In Japan, a different implementation is used in assembly operations. Instead of having a single robot perform a great number of handling and assembly tasks, the production line is serviced by many machines, each of which executes one to three motions. This process falls somewhere between a robot and hard automation. In this way the assembly can be performed by simple, inexpensive robots with a high degree of precision and few degrees of freedom. The disadvantage, of course, is the loss of flexibility built into true robots, which can quickly be converted to assembling new products.

Figure 8-10 shows an assembly station including a robot, a conveyor, a tool and gripper pallet, parts feeders, and an area for completed parts. The robot takes parts from the conveyor and the parts feeders, puts three different parts together to form an assembly, and places the assembly in the area for completed parts. The assembly is shown in Figure 8-11.

One way to improve the precision of robots with low levels of accuracy is to install a remote center compliance device (RCC) near the wrist joint. Figure 8-12 illustrates a situation in which an RCC is effective; the robot's inherent inaccuracy has brought the peg into the hole with incorrect orientation. Forces exerted at the points of contact between the peg and the walls of the hole cause motion of the gripper relative to the robot arm.

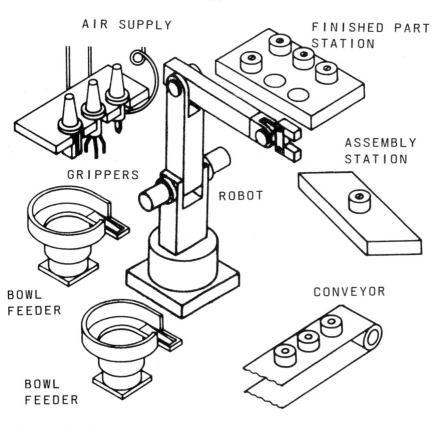

AIR SUPPLY

FINISHED PART
STATION

ASSEMBLY
STATION

GRIPPERS

ROBOT

BOWL
FEEDER

CONVEYOR

BOWL
FEEDER

FIGURE 8.10: A layout of an assembly station

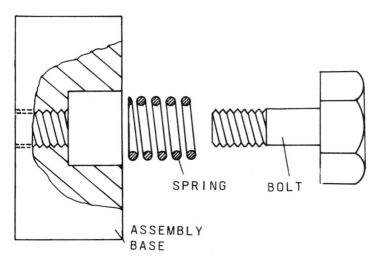

SPRING   BOLT

ASSEMBLY
BASE

FIGURE 8.11: Components to be assembled in the
assembly station shown in Figure 8.10

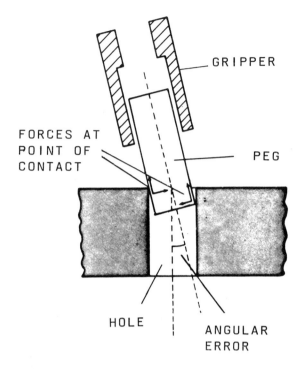

FIGURE 8.12: Compliance is needed to insert the peg into the hole effectively

Figure 8-13 shows the structure of a remote center compliance device enabling robots with low accuracy to perform assembly operations.

Another way to solve the problem of precision in assembly operations is to design the parts in such a way as to take into account the requirements of automated production. Guidelines in designed parts for automated assembly include:

* Reduction of the number of parts per assembly.

* Assembly with vertical motion only.

* Construction of symmetrical parts to reduce
  the possibility of assembly error caused by
  undesired changes in part orientation.

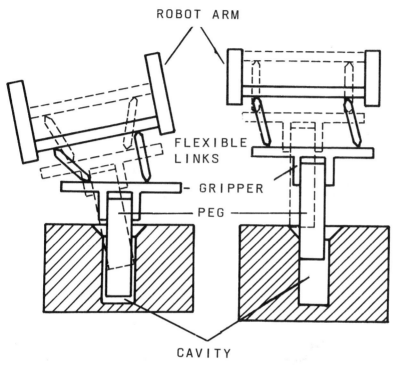

FIGURE 8.13: The structure of a remote center
compliance device

Assembly is considered a complicated application for
most robots. In fact, there are quite a few types of
assembly in which the use of robots is impractical or
impossible. A solution to this problem is provided by
the integration of humans and robots on production
lines. Robots are assigned the simpler operations;
the more complex ones are the responsibility of humans,
who also supervise the work of the robots, replacing
them in case of malfunction. To enable humans to take
over the function of the robots when a malfunction
occurs, some robot arms are designed with arms about as
long as the average human arm, so that this replacement
does not require rearrangement of the work station
equipment.

Finishing

Finishing includes the following applications:

* Sanding

* Polishing

* Grinding

* Deburring of casting

* Deburring of machining

All of these applications require the robot to move the arm along the contours of the part being machined while operating a tool at the end of the arm, or to move the part against a fixed tool. The need for constant contact between the tool and the work part means that the robot must learn many points on the surface of the part, which is a long and difficult task to teach. Therefore, continuous path control robots must be used in this application. In addition, finishing robots must have high degrees of both structural rigidity and repeatability, in order to avoid damaging parts and to provide an equal finish on all parts.

A typical finishing work station includes:

* A robot

* Tools used in the finishing process

* A unit for transfer of finished parts away from the work station

Such a typical finishing station is shown in Figure 8-14.

Changes made necessary by the introduction of automation into the work station include automatic parts feeders, used to position workpieces at desired locations and orientations. In addition, it must be remembered that the load exerted on tools operated by robots will be different from that in human operation.

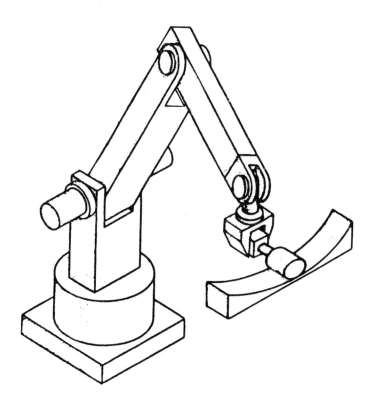

FIGURE 8.14: A robot being used for a finishing
operation

Finishing operations require the exertion of force
between the workpiece and the tool over a period of
several minutes for each part. Human operators are
limited in strength and tend to tire. Robots, on the
other hand, are capable of exerting a constant force,
which is stronger than that exerted by humans. It
follows, then, that the tools themselves must be
designed to withstand the higher work loads experienced
with robots.

Finishing is a difficult job for humans. It not only requires the exertion of continuous force, but it is monotonuous work and demands a high level of concentration. Noise, dust, and dirt cause an unpleasant environment.

To sum up, we can say that the advantages of using robots for finishing include:

* Freeing human operators from an unpleasant job.

* Improvement of product quality and uniformity, because of the elimination of the human factor.

A number of difficulties still exist in integrating robots into the finishing process. In the learning stage, the robot must be taught many points along the contours of the workpiece. In the performance stage, difficulties result from the differences between the unfinished parts and from rapid wear on the finishing machines, which distorts the dimensions of the finished products.

These difficulties may be overcome by attaching a sensor to the end effector. By sensing the contact between the workpiece and the tool, the sensor enables the robot to move the tool along the contours of the workpiece without exerting unnecessary force, and thus without damaging the workpiece.

The use of robots in finishing applications is still fairly untried, because of the demands on the finishing robot. As we have said, a high degree of mechanical rigidity and precision is required, as are continuous path control and the ability to perform complex tool motions. An important consideration is that a workpiece at the finishing stage is more valuable than at any previous stage, since work has already been done on it. Therefore, manufacturers still hesitate to use robots in finishing.

ADAPTING ROBOTS TO WORK STATIONS

In the preceding sections, we have discussed a number of examples of robot applications in industry. But, if we have left the impression that robot applications involve no more than purchasing a robot and setting it up in a work station, we must emphasize the contrary. The introduction of robots into work situations requires modifications in all aspects of the work station. In many cases, the cost of additional equipment required will be greater than that of the robot itself. To automate a work station, we need not only a robot, but also conveyors, rotary index tables, or part feeders. In addition, precision jigs are necessary to ensure that parts are properly positioned for precision work.

Proper robot operation depends on coordination between the robot actions and those of the work station equipment. This is accomplished by the transmission of signals to and from the robot. For example, a signal from a conveyor indicates that a part has been loaded into the machine, which can now commence operation. Such a signal, and the resulting coordination, are essential for tool safety, as well. For instance, a signal sent by the robot to a press might confirm that the robot has moved its arm out from under the press. Other signals used in an automated system and essential to proper operation are alerts to a damaged part, a system malfunction, or the need for raw material.

As robots become more and more sophisticated – that is, more and more like human beings – their integration into work stations is simplified. With sensors to identify part location, the jigs required for work with blind robots can be eliminated. Sophisticated robots can perform quality control inspection – can decide, for instance, whether to continue normal operations on a workpiece, repair it, or set it aside with other rejected parts and go on to the next one – all without having to alert a human operator. In effect, this is the goal of the automated work station.

Later in this chapter, we will describe the procedure for selecting a robot for a particular application and determining the changes required in the work station before the robot can be installed.

Prior to that, however, we will consider the characteristics that are relevant to the selection of a robot for a given task. They fall naturally into two categories: requisite and non-requisite, or required and not required.

Requisite Robot Characteristics

In this section, we will be discussing the work envelope, payload, degrees of freedom, precision, repeatability, and coordination with peripheral equipment.

Work Envelope

We have defined the work envelope as the collection of all points within the reach of the robot end effector. Since all operations of which the robot is capable must be performed within the work envelope, all machines and peripheral equipment to be tended by the robot must be located within this collection of points. Generally, robots are stationary and fixed to the floor. However, their work envelopes can sometimes be better utilized by hanging them from the ceiling or a wall. The work envelope can also be extended by placing the robot on a conveyor, thus making it mobile.

Payload

Payload is defined as the weight that can be carried by a robot to any point in its work envelope when it is moving at its maximum velocity. This measure is specified by the manufacturer. Engineering plans for robot applications must take into account the fact that the weight of the end effector is included in the payload.

Industrial applications of robots

Degrees of Freedom

We stated earlier that the number of degrees of freedom determines the robot's ability to execute complex motions, as well as to reach relatively inaccessible points in the work envelope.

As a general rule, a robot with six degrees of freedom can place its end effector at any point within its work envelope with any desired orientation. Obviously, this does not hold true for points whose access is cut off by obstacles, including peripheral machines. However, not all applications present such problems. Many industrial operations can use robots with only five degrees of freedom. In assembly situations requiring motion in only one direction, with a constant orientation, it is possible to achieve the desired results with robots having only three degrees of freedom.

Resolution, Precision and Repeatability

These are three of the more important characteristics in industrial robots. They relate to the deviation with which the robot places its end effector at points along its continuous path. Thus, the three characteristics are often confused. However, they are distinctly different.

Resolution is the size of the deviation of the minimum movement that the robot arm is able to undertake from any point at which it is found. The smaller the minimum deviation is, the greater the resolution, and the better it is. It is impossible to teach a robot twopoints in space where the distance between them is less than the resolution.

In robots which are taught by dictating coordinates of the points along a path, without moving the robot from point to point, the word "precision" is used to define the maximum possible deviation between any learned point and the corresponding point at the performance stage. Precision may be determined by running the robot over the path and measuring the

145

coordinates of the actual points through which it passes.

In robots taught by teach-in, teach-through, and other methods, the word "repeatability" is another measure of deviation. It defines the maximum possible deviation between any point through which the robot passed on its first performance run and that same point through which it will pass during any future series runs.

In fact, repeatability is the most important characteristic of the three being considered.

Coordination and Synchronization with Peripheral Equipment

Industrial robots are installed as components of work cells, and so they must be coordinated and synchronized (or kept in step) with the machines, conveyors, and other equipment in their environment. Robots must be able to send and receive signals to and from peripheral equipment by means of input/output (I/D) devices. In planning applications, thought must be given to the required number and position of parts feeders, conveyors, and so on, as well as to the method of coordination and synchronization between them and the robot.

Non-Requisite Robot Characteristics

In this section, we will discuss velocity of motion, type of drive, and control and teaching methods.

Velocity of Motion

A low inherent velocity usually does not prevent robots from performing tasks. A robot whose arm moves slowly can do the same work as a faster robot. However, low velocity may rule out a robot for certain applications, because it will not be cost-effective and will cause delays in production cycles.

Type of Drive

Each type of robot drive (electric, pneumatic, hydraulic) has advantages and disadvantages, as we have said. Robots with different drive systems also vary in other characteristics, such as payload, precision, and price. However, in principle, robots of all drive types can be used for all tasks.

Control and Teaching Methods

As we explained earlier, various control and teaching methods exist for robots. They allow various levels of performance and determine the degree of difficulty involved in robot operation. Less sophisticated methods of control and teaching generally do not prevent robots from performing tasks, but advanced control and teaching methods make robot operation easier and faster.

STAGES IN SELECTING ROBOTS FOR INDUSTRIAL APPLICATIONS

The process of selecting robots for industrial applications is illustrated by the chart in Figure 8–15.

To clarify the process shown in Figure 8–15, we will consider an assembly application like the one we discussed earlier in this chapter. See Figure 8–10 for comparison. We want to determine the characteristics required of the robot and of the peripheral equipment in carrying out an assembly operation.

This particular operation is composed of the following steps:

* Gripping the spring and inserting it into place.

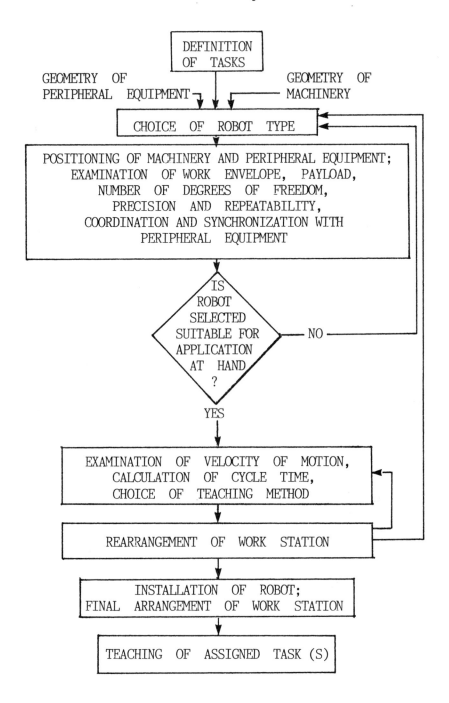

FIGURE 8.15: The process of selection of a robot for an industrial application

*   Gripping the screw and inserting it into place.

*   Substituting an automatic screwdriver for the two-finger gripper.

*   Tightening the screw.

*   Substituting a three-finger gripper for the automatic screwdriver.

*   Gripping the completed assembly and transferring it to the collection area.

Now, what are the requisite robot characteristics for the performance of this task?

Work Envelope

The robot must have a work envelope enabling it to take parts from the spring feeder and the screw feeder. It must also be able to reach the conveyor, the collection area for finished assemblies, the assembly station, and the tool and gripper pallet. A defined end effector orientation is required at each of the above points.

Payload

The robot must be capable of carrying a payload greater than the maximum weight to be transferred during the operation. In this case, the maximum weight equals the weight of the finished assembly plus the three-finger gripper.

Degrees of Freedom

The number of degrees of freedom required to perform this assembly depends on the type of robot chosen and on the orientation in which the parts are to be held.

If the robot chosen is Cartesian, cylindrical, or horizontal articulated, and if the peripheral equipment can be arranged so that all the gripping operations are performed downward from above without having to change the orientation of the end effector, then the operation can be accomplished with three degrees of freedom. Spherical and vertical articulated robots will require four degrees of freedom.

Resolution, Precision and Repeatability

In this example, resolution, precision and repeatability are determined by the tolerances of the various parts, or the deviations they allow, and by the precision required to assemble the spring and to assemble and tighten the screw.

Coordination and Synchronization with Peripheral Equipment

In this operation, coordination and synchronization must be achieved between the robot, two bowl feeders, and a conveyor. To ensure proper operation, it is advisable to provide a means of information transfer between the robot and the peripheral equipment. For example, a signal might be transmitted that there are no more parts in the bowl feeder, causing the robot to stop work. In this case, the need for the robot to stop work is not based on a safety factor, but on an economic consideration – to avoid manufacturing assemblies with missing springs or main parts.

Non-Requisite Robot Characteristics

The velocity of motion and type of drive best suited for this operation cannot be specified. Their determination depends on other factors, such as cost of production, the workpiece weight and the required precision and velocity. Concerning control and teaching methods, it is preferable to select continuous path control with the capability of linear motion, and teach-in with the same capability for linear motion. These control and teaching characteristics are desirable for such tasks as assembling parts, tightening screws, and changing tools because they require linear motion and fixed orientation of the end effector.

We have now come to the end of our discussion of robot applications in industry. There is much more to be said on this subject, at another time and in another place. However, we want to leave space in this book for a topic we have mentioned often in passing, but have not treated in detail. That topic is sensors. We have left it for the end of the book, in part, because the sensors are among the least developed components of the fully functional robot, but also because, as we have noted, they are one of the prime areas for experimentation and advancement in robotics. In the next chapter, then, we will investigate the applications of sensors and the opportunities for more sophisticated robot performance in the future through their advanced design.

Chapter Nine

Sensors and Sensing

INTRODUCTION

Futurists and science fiction writers describe robots
as machines which are made in man's image and which
assume all of man's roles in production and services.
In this chapter, we will not discuss at length the
question of physical resemblance to the human body.
However, if robots are ever to replace humans in
a wide range of jobs, considerable advances must be
made in two areas which can be compared to human
capabilities:

* Development of sensors – that is, the increased ability of robots to read information
  from their environment and increased self-
  awareness of their own characteristics.

* Development of articical intelligence – that is, a vast enlargement of the capacity of robots for comprehension, understanding, and decision. The target is for robots to be able to analyze information input from the sensors, to reach decisions based on that information, and to learn from past experience.

ARTIFICIAL INTELLIGENCE

Intelligence is the ability to understand, to know, and to learn. This ability, in humans, has made it possible for us to build and control the world, while animals having less intelligence continue to live as they have for millions of years.

Machines defined as hard automation do not have intelligence, nor do robots with limited flexibility. Some robots – those currently classified as extremely advanced – can make choices that we might call decisions, based on data gathered from their surroundings, but this level of intelligence compared to humans is very low.

Adaptation of robots to work in a changing environment is carried out by means of sensors and a processor integrated into the robot system. Sensors can be quite simple – for example, switches that signal certain events when closed or open (on/off switches); or they can be quite advanced – such as sensors that identify part types and locations in three dimensions, or measure the amount of force exerted on the robot.

However, in spite of the great development that has taken place in this field, mankind is still unable to construct a system remotely comparable to the human data entry system, which transfers information along millions of channels to the brain. Nor can we build a memory and data processing system that approaches the capability housed in the human cranium.

HUMAN VERSUS ARTIFICIAL SENSES

Some of the sensors now in existence parallel the human senses, particularly sight, touch, and hearing. Others are without parallel in the human body. Some significant examples of these unique sensor capabilities are:

*   Infrared light sensors for identification of heat sources.

*   Proximity sensors for determining the proximity of items in the area of the sensor.

*   Acoustic sensors for identification of location and motion, found in some non-human species, like dolphins and bats.

Of man's five senses - sight, touch, hearing, smell, and taste - those whose artificial parallels have been most extensively developed are sight and touch. Sight is man's principal method of data input; the quantities of information collected are truly vast. Much industrial research and development has been expended in attempts to duplicate this capability. Great efforts have also been invested in the development of touch sensors. In addition, some advance has been made in the area of hearing sensors; for example, voice identification is now possible in a few applications.

In addition to his five basic senses, which identify certain characteristics of events within his surroundings, man has the capacity to feel and identify less tangible forces and processes in the environment, such as acceleration, pressure, angle, and angular velocity. An obvious example of the use of these "sensors" is man's ability to stand erect. An inanimate structure standing on two legs is not stable, and will fall if disturbed; however, by the use of "sensors" to maintain awareness of the position of his body and legs, as well as his angle and angular velocity relative to the ground, man can stand and walk on his two legs.

The artificial sensor that compares to the human eye is the camera. Like the eye, the camera includes a lens, a shutter, and a system of detectors capable of transmitting data on the amount of light falling on them.

Artificial touch sensors consist of a "skin" that can, like human skin, send signals concerning the pressure exerted on it at any given point. Sensors can also measure forces and moments at the points of contact.

Artificial hearing sensors are <u>microphones</u>, which translate the vibrations created in the air by speech into electrical signals. In all of these instances, computers <u>decode</u> the information provided by the signals.

To date, artificial smell and taste sensors have not been developed for application in robotics.

SENSOR TYPES

Sensors now in use or under development can be classified according to the physical principle on which they are based (optical, acoustic, and so on) or according to the quantitites measured (distance, force, and so on). However, it is customary to divide them into two main types: <u>contact sensors</u> and <u>non-contact sensors</u>.

Following is a detailed listing of the kinds of information available from each of these two sensor types.

Information from contact sensors includes:

* Presence or absence of an object at a certain location

*   Gripping force

*   Force and moment exerted on the part

*   Pressure

*   Slipping between gripper and part

*   Pattern (now in the initial stages of development)

Information from non-contact sensors includes:

*   Presence or absence of an object at a certain location

*   Distances

*   Motions

*   Locations of objects

*   Orientation of objects

*   Pattern

In addition to these devices, we should mention sensor-related instruments that identify a robot's internal conditions, such as motor current sensors, encoders identifying link positions, and tachometers identifying link speeds.

In the next sections, we will discuss the principles of operation and main applications of contact and non-contact sensors.

Contact Sensors

Included in this category are those sensors requiring physical contact with the objects in their environment in order to produce a measuring signal. Contact sensors exist at various levels of sophistication. Such simple sensors as microswitches are used to identify the presence or absence of an object. Such complex sensors as artificial skins contain hundreds of sensing elements and transmit information about object orientation, dimensions, pressure exerted, and so on.)

Contact sensors, by nature, begin to supply data only after physical contact has been made between the robot and its surroundings. The contact must be made in a controlled manner; the robot arm must be moved to the contact zone slowly and carefully to avoid damage to the sensor.

Contact sensors can identify dimensions only within a volume the size of the sensor itself. In addition, they must be moved to the location where the measurement is to take place. Their main advantage lies in the precision of their measurements.

Contact sensors can be divided into two categories, according to their location on the robot arm:

* Sensors located at the contact points themselves. They permit measurement of pressure, presence of an object, identification of shapes, and so on.

* Sensors located at the robot wrist or fingers. They permit measurement of forces and moments, but not direct measurement of processes taking place at the points of contact.

Contact sensors can be further broken down into classes: single-contact sensors, multiple-contact sensing surfaces, contact sheets, slipping sensors, whiskers, and force and moment sensors. We will discuss each of these classes in the following section.

Single-Contact Sensors

The simplest contact sensors are those allowing measurement in one axis and transmitting only two possible pieces of information:

158

* "Contact exists between the sensor and some object."

* "No contact exists."

The arm controller uses this data to branch to the appropriate sub-routine, or support procedure. Figure 9-1 illustrates a two-position sensor switch installed in a gripper. This sensor tells the controller whether or not the gripper is holding an object. The controller can tell by that data whether production is proceeding normally or whether parts are no longer being fed to the robot. If parts are not being fed, the controller stops work and signals a malfunction.

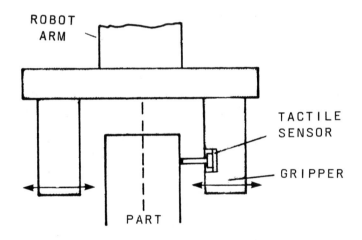

FIGURE 9.1: A tactile sensor used for identifying the presence or absence of an object

These simple sensors may be used for more complicated operations than merely determining the presence or absence of an object at a certain point. For example, they can be used to sort objects by size. This is done by gradually closing the gripper until the sensor switch is activated. By measuring the distance travelled by the gripper fingers from full open position to the point at which the switch is activated, the size of the object can be determined. Using this information, the controller branches to the appropriate sub-routine for transferring the object to the collection point for objects of its size.

This type of sensor is commonly used in automated systems, since it is simple, inexpensive, reliable, and able to supply vital data.

Multiple Contact Sensing Surfaces

A multiple-contact sensing surface is a combination of a number of single-contact sensors located in dense concentration on a single surface. Each of the sensors involved can supply an electrical signal proportional to the amount of force exerted on it. When a workpiece is placed on a contact sensing surface, all sensors in contact with the part send signals to a central processor, giving an approximate picture of the part. This process is shown in Figure 9-2.

Contact sensing surfaces can supply a great amount of information. They may be used to identify shapes of objects, to confirm the location and orientation of objects, and to compare the actual shape of an object with a desired reference shape.

Signals received from the surface are processed and transmitted to the robot controller, which uses them to decide where to move the arm in order to carry out the desired operation. Contact sensing surfaces may be used to perform quality control inspection, to sort objects by size, or to detect burrs remaining after casting. Generally, the accuracy of these procedures depends on the density or concentration of the sensors.

Contact sensing surfaces can be fixed at any point in the work envelope, or may be attached to the end effector so that they move with the robot arm. Sensors attached to the robot arm, as shown in Figure 9-3, can supply less information about the shape of the part in contact with the sensor,

160

but information about the position of the part relative to the end effector is more precise.

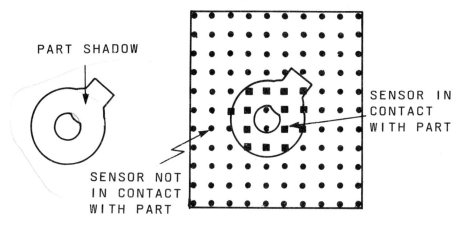

PART SHADOW

SENSOR IN
CONTACT
WITH PART

SENSOR NOT
IN CONTACT
WITH PART

FIGURE 9.2: A multiple contact sensing surface

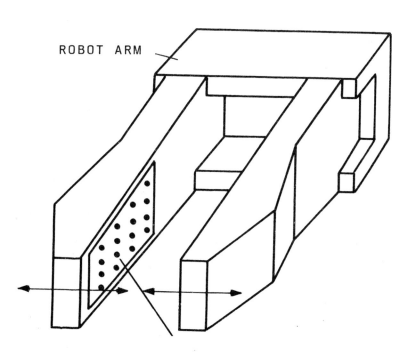

ROBOT ARM

TACTILE SENSORS FITTED TO GRIPPER JAWS

FIGURE 9.3: A contact sensing surface installed in a gripper

As we have said, most contact sensing surfaces have consisted of collections of individual contact sensors like switches or miniature load-measuring devices. Of necessity, the density of these sensors is rather low.

Recognition by means of contact sensing surfaces is still in the early stages of development, and a number of basic problems remain to be solved. These include the following:

* Physical size of current sensors.

* Distortion caused by coupling, or common wiring, between the readings of adjacent sensors.

* The cumbersome number of sensors currently required.

* Shock damage of sensing surfaces coming into contact with workpieces, as well as a loss of sensitivity in the sensing elements if they are made more rigid.

Contact Sheets

This kind of sensor can be used in situations where precise information at the contact point between the robot and the object is not required - that is, where the only need is to confirm a collision between the robot and an object in the environment. The sensor consists of a sheet of flexible material, which changes its resistance when pressure is applied to it. Contact sheets are designed basically to overcome potential safety hazards. They can be attached to any robot arm surface, and at any location that will signal the robot when its arm makes an undesired contact.

Slipping Sensors

Robots holding fragile objects must grip them lightly enough to avoid damage to the object, but tightly enough to keep it from slipping out of the gripper.

162

This fine line between too much force and too little is measured by a sensor that identifies any motion of the gripped object relative to the gripper. On receipt of a signal of such motion from the slipping sensor, the gripper automatically increases its force very gently until the motion stops.

One type of slipping sensor is composed of multiple-sensing surfaces. Any movement of the object relative to the gripper is signalled to the controller.

Another type involves a cylinder mounted inside the gripper jaws. Any slipping of the object results in rotation of the cylinder and an automatic signal to the controller.

Slipping sensors must be able to detect not only the slipping motion, but the location of the object after the slip has taken place. This information helps the robot "know" the exact position and orientation of the slipped object, so that it can continue to operate without damaging the object.

In all applications using slipping sensors, robot acceleration and deceleration must be taken into account. The grip must be firm enough to hold the object even under the most extreme changes in speed.

Whiskers

The name of this sensor indicates its mode of operation. Sensing whiskers are thin rods protruding from the end effector. Like the whiskers of a cat, they signal contact with any object in the environment. Figure 9-4 shows a sensor of this kind in operation.

Contact with an external object moves the whisker, causing the transmission of an electrical signal. The controller then gives appropriate instructions to the robot arm.

These sensors are extremely delicate and sensitive to shock. Therefore, they frequently malfunction. However, they have a number of practical applications.

163

For example, they can be used to measure the contours and surfaces of objects.

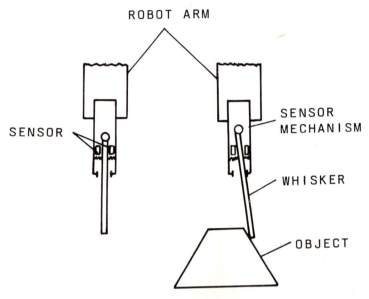

FIGURE 9.4: Sensing whiskers fitted to a robot arm or gripper will enable objects to be detected

Another use of whisker sensors is in seam tracking, a process common in arc welding. The whisker moves along the seam ahead of the welder, and the encoder installed in the whisker reports to the controller any deviation from the desired path. This application is illustrated in Figure 9-5.

To ensure correct, precise data, the whisker cannot be located very far ahead of the welding robot, because the welding process produces heat which can damage the whisker. This problem can only be overcome by manufacturing whiskers of material capable of withstanding high temperatures.

Force and Moment Sensors

The measurement of forces and moments is a common operation in many areas of engineering. As a result, a great many types of force and moment sensors exist, which utilize tested, reliable methods. It is only natural that the know-how accumulated in this field should be applied to robotics, as well. In fact, force and moment sensors are among the most common robot sensors.

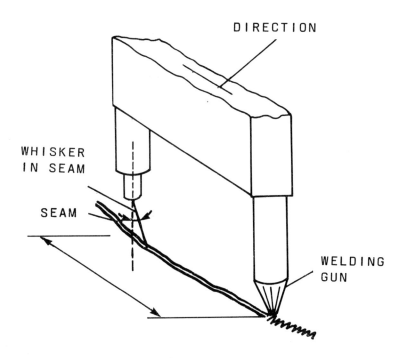

FIGURE 9.5: The use of whisker sensors for seam tracking

Forces and moments are actually measured relative to the points at which the sensors are attached, and not relative to the end effector contact points with the environment. As a result, additional processing of the information received from the sensors is required, as will be shown later in this chapter. In general, force and moment sensors are mounted between the last link of the robot arm (the link farthest from the robot base) and the gripper or tool, as shown in Figure 9-6. However, in some cases, these sensors are mounted inside the gripper fingers, as shown in Figure 9-7.

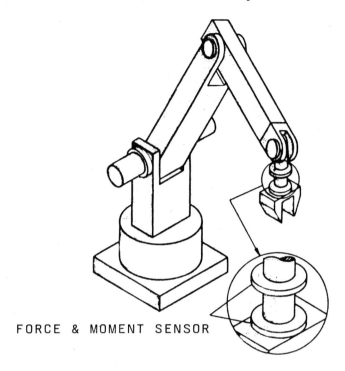

FORCE & MOMENT SENSOR

FIGURE 9.6: A force and moment sensor mounted at the joint between the last link of the robot arm and the gripper

The advantages and disadvantages of each of the methods illustrated in Figures 9-6 and 9-7 will be discussed later.

The most common method of measuring forces and moments involves measurement of the alteration in shape caused by the exertion of force. This information is taken by strain gauges, which are small pieces of conductive material glued to the object on which the forces and moments act.

Objects tend to deform slightly when forces are exerted on them. The change in shape is called strain. The strain gauge undergoes a change in shape identical to that of the object under test. Units of electrical resistance within the gauge are reported to the controller for processing and send appropriate signals to the robot.

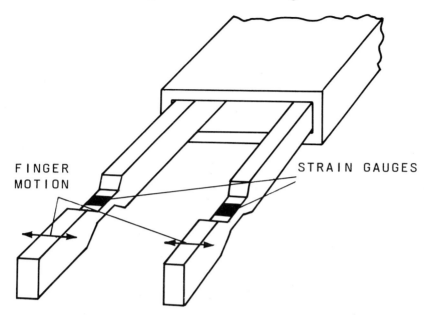

FIGURE 9.7: Force and moment sensors (strain gauges) mounted inside a robot gripper

Figure 9–8A shows a strain gauge glued to a beam, one end of which is attached to a wall. When a force, F, acts on the beam, as shown in Figure 9–8B, the beam is deformed, and the strain gauge is deformed along with it. As a result, the gauge changes its length and its electrical resistance.

The relationship between force F and the change in resistance can be described by the following equation:

$$\Delta R = C \times F \times L$$

The symbols in this equation are defined as follows:

ΔR = change in resistance

F = force acting on the beam

L = distance between the line of force acting on the beam and the strain gauge

C = constant coefficient

167

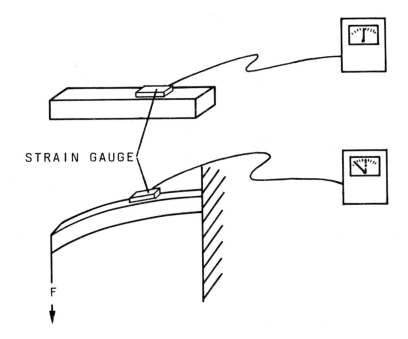

FIGURE 9.8: The reaction of a strain gauge to a load

The product of F and L (force times distance) describes the moment at the location of the strain gauge as a result of force F. Therefore, this type of equation (in which F is specified) will be used only when L is known. However, if a pure moment is applied to the beam, the sensor will identify the moment independently of the point on the beam to which the moment is applied. The equation in this case takes the following form:

$$\Delta R = C \times M$$

The symbols in this equation have the same definitions as before, expect for the following:

M = moment at the location of the strain gauge

In order to measure both force and moment acting on the beam, an additional strain gauge must be added, as shown in Figure 9-9.

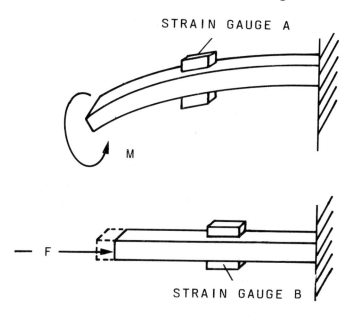

STRAIN GAUGE A

M

F

STRAIN GAUGE B

FIGURE 9.9: The reaction of a beam to a moment and an axial force

The moment M causes the beam to bend in such a way that its upper portion is extended and its lower portion is compressed. The changes in resistance noted in the strain gauges will be just the opposite. By contrast, force F will cause pressure on the beam, which will therefore change the resistance of both of the strain gauges.

The force and the moment acting on the beam may be calculated according to the following equations:

$$\Delta R_A + R_B = C_1 \times F$$

$$\Delta R_A - R_B = C_2 \times M$$

The symbols in these equations are defined as follows:

ΔR_A = change in resistance of gauge A

ΔR_B = change in resistance of gauge B

$C_1; C_2$ = defined constant coefficients

The values ΔR_A and ΔR_B are read from the strain gauges; F and M may be derived from the two equations above respectively.

To measure the forces and moments acting on other planes, it is necessary to make use of other beams with strain gauges in additional directions. In this way, it is possible to construct a sensor composed of several beams and capable of measuring all six possible force and moment components – three forces on three axes and three moments around three axes. One way of measuring forces and moments in several planes is illustrated in Figure 9–10.

GRIPPER ATTACHMENT

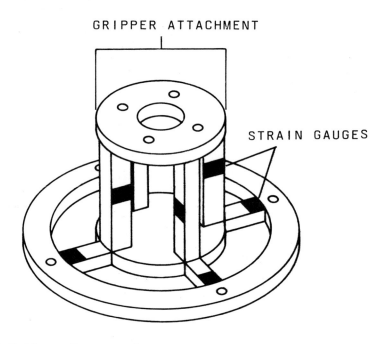

STRAIN GAUGES

FIGURE 9.10: A force and moment sensor with three force and three moment components

This sensor bases all its measurements on the moments created by forces, and not on direct measurement of these forces. This is due to the fact that this method offers higher sensitivity of measurement; in other words, smaller changes can be seen when measuring the moments of forces rather than the forces themselves.

The fingers of robot grippers may be considered as beams attached to the robot arm, in which strain gauges may be glued. In this way, a force sensor is created within the robot gripper; the force it senses is the weight of the object being held by the gripper.

The force acting on the gripper fingers in a vertical direction — that is, the weight of the object held — may be obtained from the following equation.:

$$F = \frac{R}{L \times C}$$

Here, too, the term L is used to designate the distance between the location of the strain gauge and the point at which the object is gripped. Note that L changes each time a different part is picked up. To avoid having to measure the distance when each new object is gripped, an additional pair of strain gauges can be glued in, as shown in Figure 9-11.

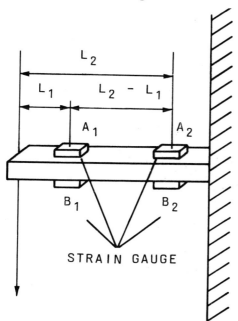

FIGURE 9.11: The use of strain gauges to measure the force applied to a beam, irrespective of the point at which it is applied

The following equations present a way to find force F acting on the robot grippers, no matter what the distance is between the gripping point and the strain gauges.

The moment created at point 1 by the application of force F is defined as follows:

$$M_1 = F \times L_1$$

The moment created at point 2 by the application of force F is defined as follows:

$$M_2 = F \times L_2$$

F may be derived from these equations in the following manner:

$$F = \frac{M_2 - M_1}{L_2 - L_1}$$

The value $L_2 - L_1$ is constant, equal to the distance between the pairs of strain gauges. The values of moment at points 1 and 2 are calculated by equations presented earlier. From the last equation presented, we see that the value of force F acting on the gripper fingers may be determined without precise measurement of the location at which the force is applied.

The last equation presented above proves that it is possible, by means of four strain gauges, to determine the weight of the object being held in the robot gripper, with the only required information being the change in conductivity of the strain gauges.

Force and Moment Sensor Applications

Up to this point, we have chiefly concentrated on the structure of force and moment sensors. We will now describe the applications of these sensors in controlling robot motions.

We will use the example of tightening screws. This is a very common and very monotonous operation. Most human operators would be only too happy to assign it to robots. However, since this operation requires the activation of force and moment "sensors" when performed by humans, it follows that the operation can be reliably performed by robots only if they are equipped with force and moment sensors.

The use of a screw to join two plates is shown in Figure 9–12.

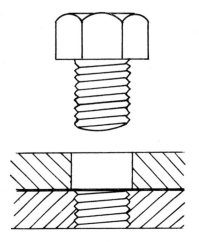

FIGURE 9.12: The use of a screw to join two plates

The operations involved in joining the plates are:

*   Gripping the screw. The sensor is used to determine that the screw has been gripped.

*   Positioning the screw. A sensor is used to determine that the screw is properly positioned.

*   Tightening the screw. A sensor is used to exert constant force on the screw, in the direction of the screw pitch.

> * Stopping the tightening of the screw. A sensor is used to sense the end of the screw course.

In the first operation, the robot uses the socket of the automatic screwdriver to hold the screw head. The force sensor identifies the added weight of the screw — given, of course, that the weight of the screw is greater than the sensing threshold, or designed lower limit. The sensor then sends an appropriate signal to the arm controller computer to proceed to the next operation.

In the second operation, the robot attempts to position the screw inside the hole in the upper plate. The robot is not expected to deal with strong forces acting on the screw at this stage. If a signal is received from the sensor indicating that a strong force is acting on the screw, the robot stops work and the controller branches to a recovery program.

The exertion of a strong force on the screw can be due to any of the following circumstances:

> * The hole in the upper plate is in the wrong place or does not exist at all.

> * The hole is too small or the screw is too large.

> * The gripper is in the wrong place.

In the third operation, the screw is tightened. The automatic screwdriver exerts a constant force on the head of the screw, in the direction of the pitch. The force required to keep the screw steadily advancing is adjusted by signals between the sensor and the robot arm.

In the fourth operation, the tightening of the screw is stopped. When the screw reaches the end of its course, the sensor reports an increase in the moment of the screw to the controller. This moment sensor is also responsible for spotting malfunctions where the screw is not tightened because the screw diameter does not correspond to the hole diameter. This is done by comparing the actual and the desired moments.

Use of Force and Moment Sensors in Finishing Applications

Figure 9-13 shows a robot sanding a workpiece. The sanding process is controlled by a force and moment sensor.

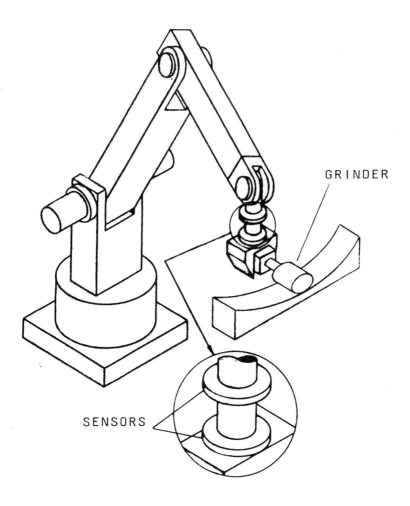

GRINDER

SENSORS

FIGURE 9.13: A robot controlled by a force and moment sensors being used in a finishing operation

This operation involves the exertion of a constant force on the sander so that it will remain in contact with the workpiece. In an earlier chapter, a sanding operation was used to illustrate the use of robots without sensors. The lack of sensors involved two problems:

* The need to teach the robot an extremely precise path, which requires much time and effort.

* The fact that the dimensions of the workpiece are reduced during – and because of – the sanding operation. Since the forces affecting the sander and the workpiece are not constant, a great many workpieces are damaged. Others may be turned out with dimensions which vary from time to time.

Adding sensors for this operation allows the system to keep track of the forces exerted between the sander and the workpiece and to instruct the robot arm to change the amount of force exerted on the workpiece whenever the measured force deviates from allowable limits. The use of a sensor makes extreme precision unnecessary in the learning stage. The path is learned approximately; the force sensor provides refinements in the performance stage as to how close the sander should come to the workpiece and how much force it should exert.

Problems in the Use of Force and Moment Sensors

An undesired effect of a force and moment sensor is that, besides measuring forces and moments at the point to which it is attached, it also measures undesired values. A sensor attached between the gripper and the robot arm measures forces resulting from the weight of the gripper itself. Therefore, it has trouble identifying a difference of 5 grams weight in the payload, if the gripper weighs 500 grams. This problem is partially solved by installing the sensors on the gripper fingers, closer to the object held.

Another problem involves the fact that the strain gauges are exposed to physical damage – the reason being that in order to increase sensitivity they are mounted on very thin rods. To protect them from abuse, a hard structure is built to house them and absorb shocks. Such a protective housing structure is shown in Figure 9–14.

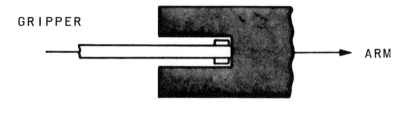

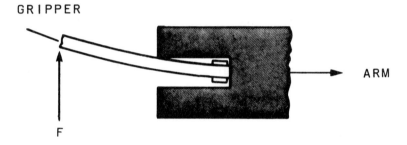

FIGURE 9.14: A protective housing for a force sensor

The development of force and moment sensors is still in the early stages, and has not progressed nearly so far, for example, as that of vision sensors. Nevertheless, the problems just mentioned have already been at least partially solved, and the use of force and moment sensors is becoming more and more widespread.

Non-Contact Sensors

Non–contact sensors, as their name implies, sense their environment without having to be brought into physical contact with the objects measured. Information on those objects is gathered from a distance. As a result, these sensors are less exposed to physical damage than are contact sensors.

The physical principle behind these sensors is the transfer of waves, which are picked up from the distant object. Media used for information transfer include sound waves, radio waves, visible and infrared light waves.

In this section, we will concentrate on light sensors, since they have undergone more development than the other types.

We will divide the identification methods used by non-contact sensors into three main groups:

* Identification at a single detector, by means of a single sensor.

* Identification along a line, by means of a sensing array.

* Identification throughout an area, by means of a camera or sensing matrix.

Identification of a Part with a Single Detector

This is the simplest type of optical sensor in existence. Its principle of operation is based on identification of a light source by means of a single detector, as shown in Figure 9-15. Under normal circumstances, the detector is exposed to rays from the light source and emits a certain electrical signal. When an object is placed between the light source and the detector, the object blocks the light. The detector reports that it can no longer see the light source, which it does by changing the electrical signal it transmits to the controller.

Figure 9-15 illustrates identification of a single point by means of a single detector. The light source and the detector are attached at opposite sides of a conveyor. When the box on the conveyor reaches the point where it blocks off the light source, the detector signals the controller that its field of vision has been blocked and activates the robot arm to grip the box and move it to another location.

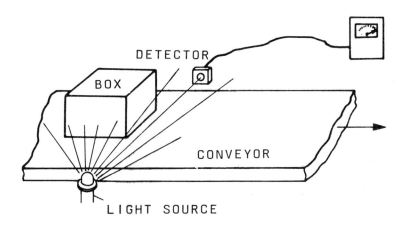

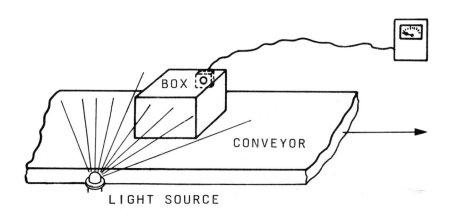

FIGURE 9.15: Identification of a point using a single detector

Single-detector sensors are very common in automated facilities, since they are reliable and inexpensive. They are generally used to identify the presence (or absence) of objects, or to identify the location of objects on a conveyor.

179

Single-detector sensors may also be used as proximity sensors. The principle of an optical proximity sensor is shown in Figure 9-16. The light source, S, sends a beam of light which falls on the object at point T. In Figure 9-16A, the reflected light from T does not reach detector D. However, when the object approaches the sensor, as shown in Figure 9-16B, the reflected light is picked up by D.

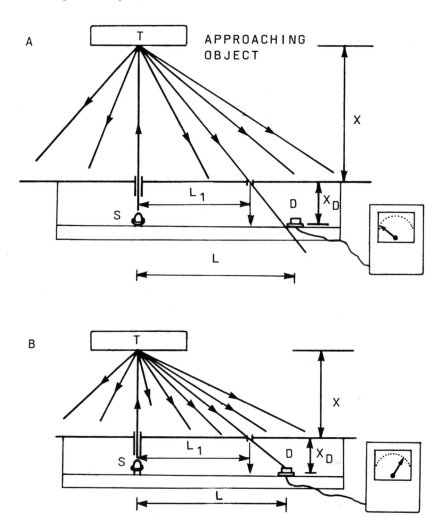

FIGURE 9.16: The principle of operation of a single-detector optical proximity sensor

The distance between the sensor and the near surface of the object may be calculated according to the following equation:

$$X = L_1 \frac{X_D}{L - L_1}$$

The symbols in this equation have the following definitions:

X = distance between the detector and the surface of the object

L = distance between the light source and the detector

L_1 = distance between the light source and the pinhole

X_D = distance between the detector and the pinhole along X

This arrangement enables efficient identification of the proximity of the arm to an object. On receipt of a signal from the sensor, the controller "knows" that the sensor has come within X distance of the object. In this application, the proximity identification serves to protect the robot arm from collision with any object.

Sensing Array (Line) Identification

An array of detectors is capable of supplying the controller with a great deal of information on the environment – much more than that supplied by a single detector. As its name indicates, a sensing array is composed of a number of detectors, arranged in a row on a common base. Detector arrays now on the market include up to 1,000 detectors and more, with the distance between detectors on the order of 10 microns (0.0004 inch).

181

The robot controller scans the detectors in the order in which they are installed, and picks up the signal transmitted by each detector in agreement with the amount of light falling on it. Identification of the presence of an object is achieved in a similar manner to that performed by a single sensor. However, the sensor array also provides other information, such as the width of the object sensed, as shown in Figure 9-17.

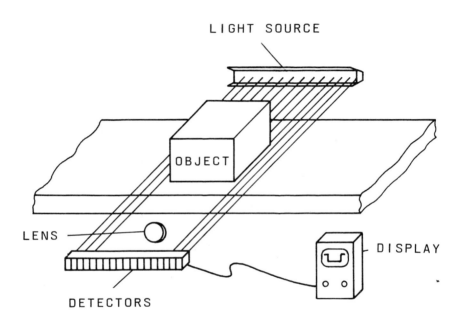

FIGURE 9.17: The use of a sensing array to identify the presence of an object and to measure it's width

Another method of identifying the presence of an object is shown in Figure 9-18. In this method, a strip of light is projected from a light source onto the sensing array. When no object is present on the surface, all of the detectors are lighted; however, when an object is present in the area struck by the strip of light, the light is not reflected at the same place. Looking at the surface from above (the direction from which the detectors see the surface), the strip of light appears to strike the object before it strikes the conveyor. In addition to identifying the presence of the object, the sensing array provides another vital piece of information – the width and location of the object on the conveyor.

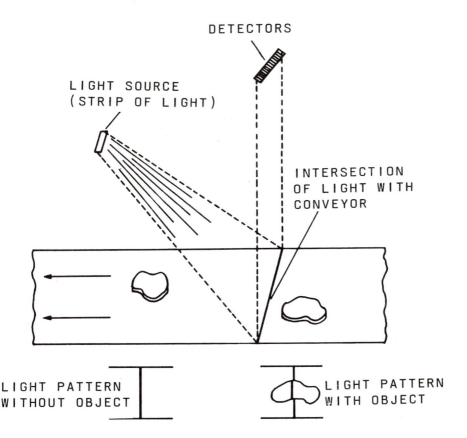

FIGURE 9.18: The use of a sensing array to identify the presence of an object moving on a conveyor belt and to measure the width of the object

If the shape of the object is simple, the sensing array supplies yet another piece of important information – the object's orientation. Figure 9–19 shows objects of varying orientation in relation to the light source. Each object blocks the lighting of a different number of detectors. By counting the number of detectors not lighted, the orientation of the object can be determined. This information can be critical when the object is to be handled by a robot.

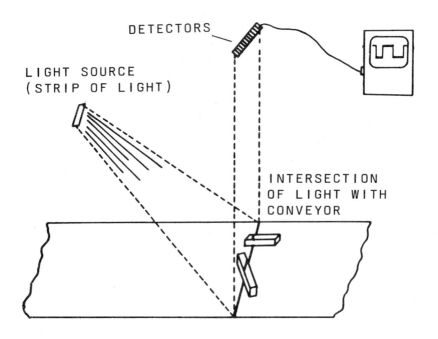

FIGURE 9.19: The use of a sensing array to determine the orientation of objects moving on a conveyor belt

Figure 9-19 shows two identical objects in different orientations. The monitor screen to the right of the illustration shows the difference in the widths of the unlighted areas generated by the two objects. Since we know the true width of both objects, we can calculate the angle of each object relative to the conveyor.

Yet another application of sensing arrays is distance measurement. The method is similar to that used to measure distance with a single sensor, as in our earlier Figure 9-16. Instead of a single detector, a sensing array is installed. As a foreign object approaches, the light transmitted from the light source is reflected onto the detectors of the sensing array. Only one detector will be lighted by the reflected light at any given instant. The change in the distance of the object from the sensing array causes the light to be reflected onto one detector after another. The distance of the approaching body is calculated with the same equation as that used to measure distance with a single sensor.

$$X = L_1 \frac{X_D}{L - L_1}$$

However, in this case, the value of L – the distance between the light source and the detector – changes according to the detector lighted at the moment of measurement. (Previously, L had a different meaning.)

This type of sensor is used for quality control of finished products. The sensor is mounted on the robot arm and passes through defined points at a certain distance from the object under test. At each of these points, the distance between the object and the sensor is measured and compared to the predetermined norm. Robots are capable of measuring hundreds of such points per hour, thus saving man from one more monotonous chore.

Limitations on the use of a sensing array attached to a robot arm include:

* Imprecision in arm orientation. This causes errors in measuring the distance, since any change in the angle between the sensor and the object under test will cause the lighting of a detector, which indicates a distance different from the actual distance at the moment of measurement.

*   Imprecision in measurement. This type of
     sensor is capable of measuring short
     distances only (up to a foot), since longer
     distances cannot be measured with precision.
     The reason for this is that the strength of
     the light reflected from the object decreases
     as the distance increases. The decrease is
     proportional to the square of the increase in
     distance – meaning that, once a certain
     distance is reached, the detectors will be
     unable to differentiate between the reflected
     light and the background lighting.

A sensing array receives information transmitted by the
reflection of a narrow beam of light. The information
contained in the beam of light relates to the line at
which the beam strikes the object. The sensing array
cannot give information relating to a surface, or to a
two-dimensional object, unless it is maneuvered in one
of the following two ways:

*   Moving the object being scanned under a fixed
     sensing array, so that the scanning affects
     the entire object.

*   Moving the sensing array over a stationary
     object.

Two-dimensional scanning to obtain information about
the structure of the object under test is generally
done with a camera. This process is discussed in the
next section.

Sensing Matrix Camera (Surface) Identification

A great deal of research has been done in the area of
camera development and identification of shapes using
cameras. Most of this effort has built on the
similarity of the camera to the human eye. This
research emphasis is reinforced by the fact that man
perceives his environment mainly through his eyes, and
most of his actions are controlled by sight data.

186

However, there are some significant differences between the camera and the human eye, mainly in the area of information processing. Methods of interaction between the human eye and the brain remain mysterious to us, and certainly cannot be imitated by the artificial means now available.

Sensing matrix cameras are constructed of two-dimensional matrices of sensors arranged densely on a common surface. The number of sensors per matrix varies from 32 x 32 (that is, 32 rows of 32 detectors each) in low-resolution sensors to 512 x 512 in high-resolution sensors. The distance between detectors is measured in tens of microns (thousandths of an inch).

The image of the object under observation is projected onto the sensing matrix by means of lenses. The detectors are electrically scanned, and a signal is obtained from each detector, proportional to the amount of light falling on it. The amount of data gathered from the camera is immense; a 512 x 512 matrix includes over 250,000 detectors, meaning that each scanning cycle yields over 250,000 bits of data.

The robot will not be able to operate on the object under test unless it is equipped with a vision sensor, to identify the nature, the precise location, and the orientation of the object seen by the camera. The number of data bits which must be transmitted to the robot to define the object and its location and orientation is far smaller than the 250,000 bits gathered by the camera. The obvious conclusion is that the information gathered by the camera must be processed before it is transmitted to the robot. This type of processing is called (pattern recognition) A considerable amount of research has been done in this field, since it is of vital importance to the military and in many areas of engineering. As a general rule, this signal processing requires an additional computer besides that of the robot controller. Also required is some pre-processing of the analog signals received from the camera. The type of identification and decoding system used in pattern recognition is shown in Figure 9-20.

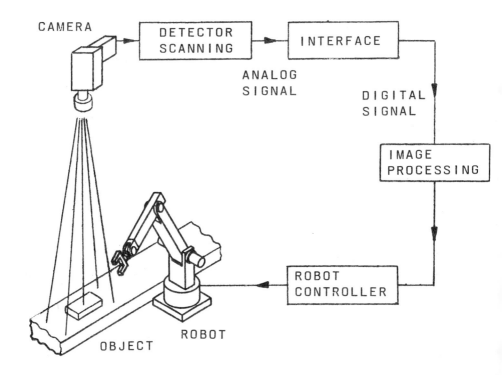

FIGURE 9.20: Processing of camera-gathered
information for use by a robot controller

To explain the exact purpose of each system component,
we will consider an example of the identification of a
cylindrical part. Since this part is symmetrical, it
does not require identification of orientation. Therefore, the system requires only the location of the
object's center of gravity. The object is lighted by a
light source. Its image is cast on the sensing matrix
by a lens. The object image on the sensing matrix is
shown in Figure 9–21A. The data transmitted by the
detectors do not form a complete circle, as shown in
Figure 9–21B. Pattern recognition of this object is
performed by comparing the area of the object
silhouette to a table of object areas stored in the

computer memory. In this case, the area of the object
silhouette can be determined by the number of sensors
covered by the object image, which is compared to a
table of numbers representing the areas of familiar
objects. See Figure 9-22.

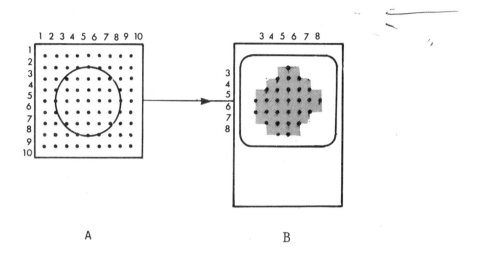

A   B

FIGURE 9.21: The silhouette of an object as
identified by a light sensor

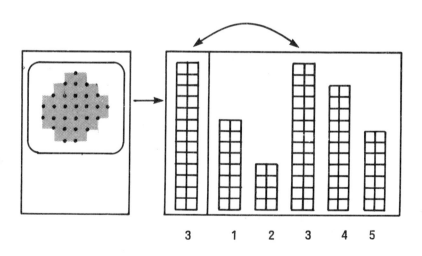

3   1   2   3   4   5

FIGURE 9.22: Object identification by comparison of
it's silhouette with an object table stored in
computer memory

189

Basic Concepts

By comparing the object area to a table of familiar object areas, the computer identifies the cylindrical object as one of the familiar objects (in our example, object 3).

At this point, the robot "knows" the identity of the object; its location, however, is still unknown to the system. Identification of the object location will be carried out by location of the center of gravity of the image cast on the sensing matrix.

The center of gravity of the object image is calculated according to the following equations:

$$X_{c.g.} = \frac{1}{N} \sum_{n=1}^{N} X_i$$

$$Y_{c.g.} = \frac{1}{N} \sum_{n=1}^{N} Y_i$$

The symbols in these equations are defined as follows:

$X_{c.g.}$ = X-coordinate of the center of gravity

$Y_{c.g.}$ = Y-coordinate of the center of gravity

N = number of detectors covered by the object image

X_i = X-coordinate of detector i ($1<i<N$)

Y_i = Y-coordinate of detector i

In this example, the object image falls on 26 detectors, meaning that N = 26. The values of X as shown in Figure 9–21 are:

$X_1 = 3$, $X_2 = 3$, $X_3 = 3$, ---- $X_{26} = 8$

The values of Y as shown in Figure 9–21 are:

$Y_1 = 3$, $Y_2 = 3$, $Y_3 = 4$, ---- $Y_{26} = 8$

Using the equations above, we find that the center of gravity of the object image cast on the detectors is located at the following point:

$$X_{c.g.} = 5.35$$

$$Y_{c.g.} = 5.65$$

Once we have located the center of gravity of the object image cast on the sensing matrix, this information must be translated in order to find the center of gravity of the actual object. This can be performed quite simply, based on the ratio of the sensing matrix area to the field of view as shown in Figure 9-23.

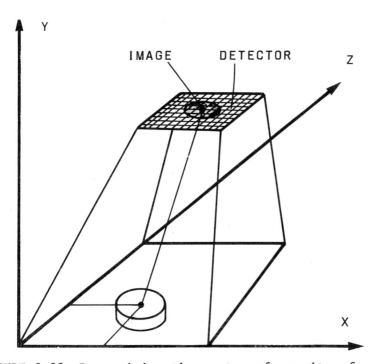

FIGURE 9.23: Determining the centre of gravity of an object using the ratio of the sensing matrix area to the field of view

The object has now been identified and its center of gravity has been located. The robot is therefore capable of performing the desired operations on the object.

In this example, the object was radially symmetrical, so its orientation did not have to be considered. However, in more complex cases, additional details may have to be identified. These may include:

* Object length and width

* Orientation of the object

* Existence and location of hollows within the object

* Moment of inertia

* Many more variables

To identify these variables for any given object, the contours of that object must first be identified. A number of algorithms have been developed for use in identifying object contours. These algorithms, collectively referred to as (edge detectors) are complicated and generally require extensive software as well as long computer processing time.

Up to now, we have described the identification of an object according to its image cast on a sensing matrix. However, a camera may also be used to identify points of light projected onto the subject, and thus to identify its shape. A simple example of this process involves using a light source to manually light points around the edges of an object. The camera identifies each point and transmits its coordinates to the computer. Since the points are all on edges, the computer can connect them, and thus arrive at a representation of the object's contours. A more complex example is the lighting of the same object with strips of light, similar to the method described above for sensing arrays.

Identification by a strip of light and a camera is used in arc welding. Since the reflected light is not continuous between the two plates to be welded, the camera identifies a break in the projected line of light. The location of the break parallels the location of the seam. In this way, the robot controller identifies the path required to weld the plates together. Figure 9-24 shows identification of a weld seam using a projected strip of light.

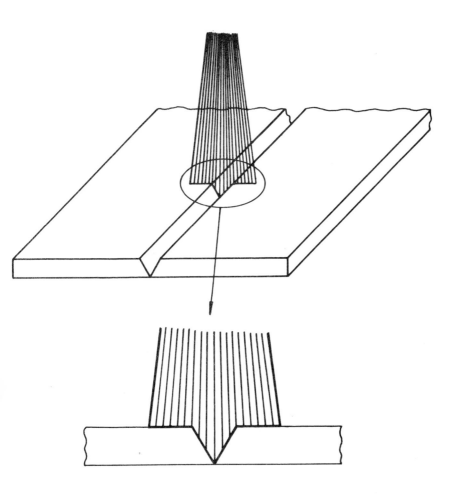

FIGURE 9.24: Identification of a weld seam using a projected strip of light

In arc welding, the vision sensor is mounted in front of the welder and includes a light source and a camera. The light source lights the seam an inch or two ahead of the welder; the camera detects the shape of the seam and identifies the proper location for the weld. The information picked up by the camera enables the robot controller to direct the robot arm and perform the welding operation.

Throughout this section, all examples of identification have been two-dimensional only - identification of the shape of the surface. Naturally, the detectors used in this type of identification are also arranged on a surface. How, then, can cameras be used for three-dimensional identification of objects?

One method, suitable for simple objects, is accomplished by projecting a number of strips of light onto the object and measuring the breaks in the reflected strips of light. The angle of the break and the extent of the broken segment serve to indicate the height of the object.

Another method, enabling three-dimensional identification of more complex objects, involves two cameras in two different locations. They look at the object stereoscopically, much in the manner of human eyes. This form of identification is complex, requiring coordination between the cameras and complicated processing of the information which was gathered by both cameras.

The method used for the identification of the cylinder and its location, as shown earlier in this section, requires binary signals from the camera. This means that all detectors covered by the object image transmit a "1" signal, and all detectors not covered by the object image transmit a "0" signal. In this manner, the computer identifies the contours of the silhouette (the image produced when the object is located between the light source and the camera). However, this method provides information on the silhouette only, and not on the internal structure of the object - such as its varying height.

To overcome this problem, researchers have developed a method in which identification is not only in black and white, but also in shades of gray, in a manner similar to the visual identification methods used by man. This method, called the gray scale, is based on the fact that different surfaces reflect different intensities of light, according to their angle relative to the light source. The camera detectors and the processor are designed to differentiate between levels of gray. Analysis of the various shades of gray enables the system to determine the lines of intersection of two surfaces. Thus, information is provided about the structure of the object examined, and not only on the contours of its silhouette.

To date, great efforts have been invested in the development of vision sensors. However, the number of sensors in active use in industry is still small. We can give several important reasons for this situation:

* Vision sensors enable identification of simple shapes only.

* Illumination of objects and reception of reflected light pose a great many problems.

* Three-dimensional identification is complicated and requires expensive equipment and cumbersome software.

* Identification is only possible when the parts are separate from each other. It is difficult to identify parts dumped into a bin and to select one of them to be pulled out. This process is called bin picking.

It appears that a breakthrough in vision sensors and image processing will still be necessary before they can be efficiently used in many industrial plants. Solving these remaining problems will allow dramatic advancement toward the comprehensive use of robots in both industry and services. The next chapter will discuss the possible "world of the future," when the breakthrough has occurred.

Chapter Ten

Future Developments

PREVIEW

Most, if not all, human beings eagerly await the day
when machines will take over all the difficult, tiring,
and boring jobs. Is there really anyone who would
not want a devoted, loyal mechanical servant, ready
to obey any and all orders, exactly and
unquestionably?

It appears that a revolution in the field of robotics
is upon us. The vast resources devoted to robotics
research and development have already begun to bear
fruit, and will certainly lead to even greater
achievement in the future.

A great many tools have already been developed, and
many energy sources have been exploited, to replace

humans and assist them in their labor. To date, however, man has remained an important part of the system. Admittedly, he no longer has to use his muscles as strenuously as in days gone by; he does, however, use his senses – and, above all, his intelligence – to analyze situations and make decisions. An example of man–machine combinations may be found in the operation of a bulldozer. Man no longer has to dig or uproot stumps and stones. But he is responsible for deciding where to drive the bulldozer and for checking to see whether it has carried out its assigned tasks.

The main aim of robotics development today is to release man from any contact with hazardous, difficult, and boring tasks. To achieve that goal, robot designers must give the machines the ability to identify their environment and to make decisions.

(a) To replace humans with robots, future research will focus on the development of artificial intelligence. At the same time, research will be conducted on the human brain, in order to understand how this marvellous mechanism gives man such high intelligence.

Robotics is now in its early stages of development. It is still considered by many to be a disorderly mingling of many fields – control, computers, artificial intelligence, and so on. There is still no one, clearly defined, self–contained field of robotics.

This situation could well change in the future, however, as occurred in the field of computers. Originally considered a branch of electronics, computers now represent a separate discipline entirely. This came about as a result of the immense body of knowledge accumulated in the field. A would–be computer expert does not have to learn electronics from the ground up; his studies are supported by existing electronic systems, which he is not required to learn in great detail.

The emergence of robots is basically a result of the immense development in the area of computers. The next step, however, is dependent on the appearance of computers which are much more sophisticated than any we have known to date. That is, computers are needed that are capable of serving as robot brains.

How do we envision the robot brain of the future?

The robot brain will be a sophisticated computer which will receive information from robot sensors, process it rapidly and efficiently, and control robot motions. It will have to include a large memory and a capacity for analyzing complicated situations. It will make all decisions regarding the robot's course of operation. So as not to interfere with mobility, the robot brain will have to be small in size and extremely low in power consumption.

UNMANNED FACTORIES

In industrial circles today, a prime objective is the development of plants whose production lines will not require the presence of any human beings. Such installations are called unmanned factories.

In the unmanned factory of the future, the entire production line will be controlled by computers, which are connected to machines, robots, and sensors. Robots will perform tasks in production and in tending automated machines. Sensors will be used in quality control and supervision of the production process.

The central computer will be called upon to perform process planning, as well as to identify malfunctions and to overcome them. Changes in the production line – such as switch-over to a new product - will require minimal human assistance. The central computer will receive product orders from marketing departments (or directly from customers) and will decide which equipment to operate and when, in order to meet production goals.

The unmanned factory will be able to work continuously, including nights and weekends. It will require no lighting except for a central control room, in which several humans will keep track of system functioning for the entire factory.

The first steps toward realizing this vision have already been taken in Japan. The Japanese government has invested tremendous sums in the advancement of robotics and automated production systems, in order to proceed to the goal of the unmanned factory.

PERSONAL ROBOTS

A personal robot is an automated machine whose operation is not in the field of industrial production, but in services to mankind - in private homes, restaurants, and so on. Its development represents a challenge which is far from solved; in fact, it is doubtful whether a fully functioning personal robot will be achieved in the near future.

The potential market for personal robots is nearly unlimited - like the market for automobiles, television sets, and other consumer goods. The necessary effort has not yet been invested in the development of personal robots for the following reasons:

* In today's technology, the price of a personal robot would be very high.

* Performance requirements for personal robots are more stringent than those for industrial robots - more sophistication, less routine, more mobility.

* The safety factor is more critical. A robot in an industrial plant is generally

surrounded by a safety barrier or fence to prevent the uncontrolled approach of humans while the robot is in operation; opening the barrier causes an immediate stoppage of work. By contrast, personal robots would have to work in crowds of human beings. It would therefore have to be much more reliable, so as not to injure humans if it should malfunction.

* A prerequisite for the use of robots in industry – or in private homes – is the ability to perform operations with precision. Personal robots designed according to today's technology would suffer from limited precision, because of the need for mobility. Only an intelligent robot, capable of sensing its environment and analyzing existing situations, ..ould be able to overcome the problem of precision.

MAN-MACHINE COMMUNICATIONS

Great leaps forward are expected in the area of man-robot communications. Today, robots are programmed to keep performing the same operations for long period of time. Therefore, programming by keyboard is accepted practice, since it is rapidly accomplished, compared to the period of time the robot will be operating under the program. In the future, with increased flexibility in transferring from one operation to another, robots will be used not only in industry, but in private homes; however, by then, keyboard programming will be considered much too slow.

It seems certain that the preferred method of man-machine communications in the future will be by voice. Therefore, much effort will have to be invested in advancing the area of speech recognition and analysis of information conveyed by voice.

ARTIFICIAL BODIES

As of today, the development of prostheses – artificial limbs and organs – is not considered part of robotics. Its main goal is the development of substitutes for injured parts of the human body – arms, legs, internal organs. These prostheses are made of lightweight, sturdy, reliable materials. As such, the technology involved can be utilized for the future development of sophisticated, nearly humanoid robots.

CONCLUSION

The future holds in store a nearly unimaginable potential for advancement in the field of "smart robots." In the distant past, most of mankind labored in the production of food. The Industrial Revolution, with its multiple effects on daily life, changed society beyond recall.

The development of smart robots will bring about another revolution, no less crucial, in human life. Machines will be responsible for production and service jobs, and man will be free to pursue scientific research, academic professions, hobbies, and relaxation.

Glossary

The first number following the definition in each entry of the glossary serves as an index to the first occurrence of the term. Each following number refers to a page on which a significant discussion of the term occurs.

AC MOTOR: a motor that operates on alternating current. 47

ACOUSTIC SENSOR: a type of sensor that identifies location and motion based on analysing acoustic waves. 155

ACTUATOR: a device converting electrical, hydraulic or pneumatic energy into mechanical motion. 56

AIR GAP: an air space between the welding gun and the metal object being welded. 126

ALGORITHM: a procedure for solving a mathematical problem in a finite number of steps that frequently involves repetition of an operation. 192

AMPLIFIER: a component used to increase the power of a signal. 57

ANALOG TO DIGITAL CONVERTER (ADC):   a device that converts physical signals (like voltage or amount of motion) into their digital representation. 57

ANGULAR ERROR: lack of precision in the orientation of a part. 134

ARC WELDING: a process used to join two metal parts along a continuous contact area.   114, 126

ARM: an interconnected set of links and powered joints comprising a manipulator and supporting a moving end effector. 7, 44, 45, 46

ARRAY: overall arrangement of a machine. 12

ARTIFICIAL INTELLIGENCE: the capacity for independent thought, analysis, and decision-making performed by a computer. 15, 154

ASSEMBLY: the fitting together of manufactured parts. 114, 133

AUTOMATIC GRIPPER CHANGER: a adapter which allows a gripper to be quickly attached to the end of the robot arm and allows for a change in gripper. 102

AXIS: an hypothetical line which defines a direction. In robotics, often used as a synonym to a degree of freedom in the mechanical arm. 7

BACKLASH: a looseness between tooth wheels, gears, belts, and other transmission devices. Also called mechanical freedom. 53

BALANCE UNIT: component that supports the static weight of the arm. 86

BALL AND SOCKET JOINT: a connection that functions like a combination of three revolute joints, allowing rotary motion around three axes. 26, 29

BASE: the foundation of the robot, usually fixed to the floor, a wall, or a ceiling. 25

BASE JOINT: the joint in the mechanical arm closest to the base. 108

BINARY: a binary number is composed of a sequence of ones and zeros - in other words a series of true and false values (or an ON or OFF state). 194

BIN PICKING: a process of identifying and selecting one of the parts dumped in a bin in order to pull it out. 195

Glossary

BLIND: in a robot, having no vision sensor. 131

BOWL FEEDER: Peripheral equipment such as a hopper or bowl which feeds parts to be processed in an orderly manner from a batch of components which are randomly stored. 150

BRAIN: computer that controls robot's movement. 7

CARTESIAN: in robotics, a type of robot having three prismatic joints. 32, 33, 40

CLOSED-LOOP CONTROL: a type of control in which feedback provides continuous information to allow for correction of errors. 57, 58

COMPARATOR: compares the command and the actual position. Issues an error signal, causing the system to correct the joint motion. 60

COMPUTERIZED NUMERICAL CONTROL (CNC): a form of control in machines that are operated by computers controlling the machine motor, used mainly in machine tools. 5

CONTACT SENSOR: a sensor requiring contact with the objects in their environment in order to produce signals. 156, 157

CONTACT SHEET: a type of sensor that confirms a collision between the robot and an object in the environment. Consists of a sheet of flexible material which changes its resistance when pressure is applied to it. 158, 162

CONTINUOUS PATH CONTROL: type of control which the arm must move along an exactly defined path. 66, 67

CONTROL HIERARCHY: levels of control with each level controlling the previous one. 56

CONTROLLER: that part of the robot which operates the mechanical arm and maintains contact with its environment. 24, 55, 110

CONTROLLER SOFTWARE: programs that must be written by an experienced software expert with detailed knowledge of the controller computer. 73, 74

COORDINATES: any of the magnitudes which define a position of a point. 64, 88, 89

CYLINDRICAL OBJECT GRIPPER1 a gripper consisting of two fingers, each marked with several semi-circular indentations. May hold objects of several different diameters. 98

CYLINDRICAL ROBOT: a robot having one revolute joint and two prismatic joints. 32, 34, 41

DC MOTOR: electric motor operated by DC current. 47

DEBURRINGL smoothing off areas of roughness produced in the cutting or shaping process. 140

DECODE: to interpret. 156

DEGREES OF FREEDOM; in robotics this relates to the number of movements which are totally unrelated which the robot is able to perform. 31, 145, 149

DEVIATION: the difference between the desired position and the actual position of the end effector. 145

DIE: a form or mold into which molten metal or other material is forced. 114

DIE CASTING: the process of injecting a raw material at its melting temperature into a special form or die. 114, 116

DIGITAL: a description of data in a numerical format. 57

DIGITAL TO ANALOG CONVERTER: a device that transforms digital data into analog data. 57

DIRECT DRIVE: a type of drive where a motor is mounted directly on the joint it moves. 46, 51

DUMMY POINT (OR VIA) POINTS: are points in the path of the robot arm through which it passes without requiring gripper motion. 107

EDGE DETECTION: a collective term for the algorithms that have been developed for use in identifying object contours. 192

ELBOW JOINT1 a joint on the articulated arm that is located between the shoulder joint and the wrist (see the human elbow). 108

ELECTRIC DRIVE: a type of drive in which electric motors are connected to a source of electric current. 47

ELECTROMAGNETIC FIELD: an area surrounding an electromagnet that contains magnetic sources. 101

ELECTROMAGNETIC GRIPPER: a device that is designed to attach to ferromagnetic objects by creating an electromagnetic field. 101

ENCODER: a feedback device which provides information on the position of the links of the robot. The encoder converts position data into electrical signals. 59, 62

END EFFECTOR: a tool, gripper, or driven mechanical device attached to the end of a manipulator by which objects can be grasped or otherwise acted upon. 8, 95

ENVELOPE1 the maximum range of motion attainable by the robot in all directions. 9, 38, 144, 149

ERROR SIGNAL: the difference between programmed and actual joint status. 60

FEEDBACK: the signal fed back to the commanding device. The signal is used to compare the desired response and the actual response of the system, and to make corrections to reduce the error. 58

FERROMAGNETIC: relating to substances with an abnormally high magnetic permeability. 101

FINISHING: the final treatment of a surface. 114, 140

FLEXIBLE AUTOMATION: machines which can be easily programmed, and which can change over easily from one manufacturing setup to another. 11, 13

FORCE AND MOMENT SENSOR: a common type of sensor that measures force and moment relative to the points at which the sensor is attached. 81, 87, 158, 164, 172, 175, 176

FRAGILE OBJECT GRIPPER: a type of gripper designed to handle delicate objects. 99

GANTRY: a type of track on which a robot can be suspended to make it mobile (in general, suspended from the ceiling). 120

GENERATION: classification of evolution. In robots, according to their levels of sophistication. 14

GRAY SCALE: a method in which the camera detectors and the processor are designed to differentiate between levels of gray. 195

GRIPPER: a type of end effector used to grasp parts. 24

HARD AUTOMATION: machines which are designed to perform specific functions. Every change in standard operation demands a change in machine hardware and setup. 11

HARDWARE: the mechanical, electrical and electronic devices of which a computer is built. 55

HIGH-LEVEL LANGUAGE: a simplified computer programming language using normal human language for instructions. 74

HORIZONTAL ARTICULATED ROBOT: a robot having one prismatic and two revolute joints. 32, 36, 41

HYDRAULIC DRIVE: a type of drive unit that causes motion in the parts by compressing oil. 46, 48

INDRIECT DRIVE: type of drive in which the motor is mounted at a remote point from a joint. 46, 51

INFRARED LIGHT SENSOR: a sensor identifying heat sources by means of infrared light. 155

INPUT: an external signal to the robot, normally used to synchronise the robot and its surroundings, e.g. machine tools, conveyors, etc. 10

INPUT/OUTPUT DEVICE: a device used to send a receive signals. 146

INTELLIGENCE: the capacity for independent thought, analysis, and decision-making. 9

INTERFACE: to have the ability to communicate back and forth (in computers and robots the interfaces are used to synchronise the electronic inputs of one sort with those of another sort). 10

INTERPOLATION: computation of intermediate points to resemble a path control. 67

JIG: a device used to maintain mechanically the correct positional relationship between a workpiece and a tool. 129

JOINT: the component that allows the relative movement between two links, in a robot, human body, etc. 25

JOINTED GRIPPER: an end effector with a large number of links and joints. Designed to grasp objects of irregular size and shape. 100

JOYSTICK: a device that has switches to control robot motion. 83

KINEMATIC CONFIGURATION: collective reference to the examination of types of joints and their order. 42

LIGHT PEN: device that uses narrow beam of light as a pointer to input points into the controller's memory. 91

LINEAR: extended in a line. 27

LINK: a segment of a robot arm. Connected by joints to other segments. 25, 44

LOAD: object or material to be moved by robot arm. 7

MACHINE TOOL LOADING AND UNLOADING: loading and un- loading of machine tools that form workpieces. 114, 119

MANIPULATOR: mechanical arm consisting of a series of links, jointed or sliding relative to one another, for the purpose of grasping objects. 7

MASTER ROBOT: a lightweight machine that records motions which the controller can then use to command a "slave" robot. 85

MECHANICAL FREEDOM: a looseness between tooth wheels, gears, belts, and other transmission devices. Also called "backlash". 53

MEMORY: main storage used for temporary storage of programs, input data, and output results. 7

MICROPHONE: an artificial sensor which translates vibrations in the air caused by speech into electrical signals. 156

MICROPROCESSOR: single integrated circuit, acting like a small computer that has full computing power. 10

MICROSWITCH: a simple sensor used to identify the presence or absence of an object. 157

MINICOMPUTER: type of computer whose physical size is usually smaller than a mainframe computer. Its performance generally exceeds that of a microcomputer. 10

Glossary

MOMENT OF INERTIA: the property of matter by which it retains its state of rest or of uniform linear motion so long as it is not acted upon by an external force. 33

MONITOR SCREEN: a screen used to view a picture picked up from a camera, or text in communications with a robot. 91

MULTIPLE-CONTACT SENSING SURFACE: a combination of a number of single-contact sensors located in a dense concentration on a single surface. 158, 160

NON-CONTACT SENSOR: a device that senses its environment without having to be brought into physical contact with the object measured. 156, 177

OFF-LINE PROGRAMMING CONTROL: a mode in which the robot controller stores the path of motion in its memory as a series of points and corresponding motions of the various joints. 71

OPEN-LOOP SYSTEM: a system that does not provide feedback to the controller.   59

OPTICAL ROTARY ENCODER: a type of encoder that is composed of a light source, a light detector, and a slitted disk which revolves between the light source and the detector. 62

OPTICAL SENSOR: a device that senses visible or infrared light waves and transmits an electrical signal to the controller. 178

ORIENTATION: posture or positioning of a workpiece. 43

OUTPUT: a signal of the robot to external equipment, generally used for synchronization between the robot and the equipment, such as machine tools, conveyors, etc. 10

PALLET: a portable platform for handling, storing, or moving materials. 115

PATH CONTROL: the control of each robot arm with coordination between the axes to form the required path. 56, 65

PATTERN: a defined form or structure. 157

PATTERN RECOGNITION: a type of processing to identify an object by its pattern. 187

PAYLOAD: the load being carried by the robot. 124, 144, 149

PERIPHERAL EQUIPMENT: any unit of equipment, distinct from the central unit, which may provide the system with outside communication (in robotics conveyors, turntables, etc). 15, 146, 150

PITCH: movement of the gripper up or down. 43

PLANAR: two-dimensional in quality. 101

PNEUMATIC DRIVE: a type of drive system similar to hydraulic, except that air flows through the system instead of oil. 46, 49

POINTER: a light pen used as a pointer to input desired points on the screen to the controller. 81, 91

POINT-TO-POINT CONTROL: a type of control that defines a series of points in space for the robot arm movement. 66

POTENTIAL DIFFERENCE: the voltage difference between two points that represents the work involved or the energy released in the transfer of a unit quantity of electricity from one point to another. 126

PRESSING: an operation used in forming parts. 114

PRESS LOADING AND UNLOADING: a process of taking an unmachined part from the part feeder and placing it in the press, then transferring it from the press to a conveyor belt. 114

PRISMATIC JOINT: a type of joint composed of two nested links which slide into or beside each other.   26, 27

PRODUCTION CELL: a flexible unit which inputs raw material, machines it stage by stage, and outputs a finished product.   120

PROGRAM: a planned sequence of instructions that tells a computer system what steps to perform.   7, 71

PROGRAMMING: a method of instruction in which the robot, in the teaching stage, does not move physically from point to point, but is instructed by information in the controller's memory.   10, 81, 88, 92

PROXIMITY SENSOR: a device that senses an object only a short distance away and measures how distant it is.   155

REAL-TIME   CONTROL: control   is   effected   in   the actual   time   in   which   a   physical   event occurs.   The   results   of   calculations performed   during   this   actual   time   are used   for   the   purpose   of   continuing   the event.   71

RECORDING PUSH-BUTTON: a device which causes the controller to record in its memory the position of the joints and the status of the end effector. 82

RECOVERY PROGRAM: the process by which the controller branches to an alternate action instead of stopping the robot arm when a sensor signals a problem.   174

REMOTE CENTER COMPLIANCE DEVICE (RCC):   a device used to interface a robot or other mechanical workhead to its tool or working medium.   This enables a robot which has a low accuracy to perform assembly work which requires a higher degree of accuracy.   136

REPEATABILITY: the radius of the sphere around any previously learned point within which the robot can locate its tool center point. 127, 145, 150

RESOLUTION: the least interval between two adjacent discrete details, or points, that can be distinguished from one another. 187

REVOLUTE JOINT: a connection that permits rotary motion between two links. 26, 28

ROBOT: a mechanical arm, or manipulator, designed to perform many different tasks and capable of repeated variable programming. 7, 105, 143

ROLL: rotation of the gripper. 43

ROTARY: moving in a circular direction. 28

ROTARY INDEX TABLE: a device used in the precise positioning of workparts. 129

SAMPLING: readings taken at fixed intervals of time. 85

SEAM   TRACKING: a process that uses a sensor to track a seam. In robotics, applied in arc welding. 164

SENSING ARRAY: a number of detectors arranged in a row on a common base. 181

SENSING MATRIX CAMERA: a two dimensional matrix of vision sensors arranged densely on a common surface. 187

SENSING THRESHOLD: the designed lower limit of the force and moment sensor. 174

SENSOR: a device which supplies information on the environment. 9, 153

SERIES: a collection of steps which combine to form the robot's operating program. 66

Glossary

SHIFTING COORDINATE SYSTEM: a programming method in which the robot is taught identical paths for tasks in two or more work cells. 89

SHOULDER JOINT: the joint on a vertical articulated mechanical arm that is located between the base joint and the elbow joint. 108

SILHOUETTE: the outline of a body viewed as circumscribing a mass. 188

SINGLE-CONTACT SENSOR: a simple sensor that allows measurement in one dimension and transmits only two possible pieces of information. 158

SLAVE ROBOT: the robot that carries out the task taught by a "master" robot. 85

SLIPPING SENSOR: a sensor that identifies any motion of the gripped object relative to the gripper. 158, 162

SOFTWARE: programs that instruct the operations of a computer. 10

SPHERICAL ROBOT: a robot with two revolute joints and one prismatic joint. 32, 35, 41

SPOT WELDING: a process in which a high current flows between two electrodes and through two pieces of metal which are to be joined. 114, 121

SPRAY PAINTING: in robotics, a process involving the attachment of a spray gun as an end effector. 114, 130

STEP: a part of the robot's program. 66

STEPPER MOTOR: a sub-class of DC motors moved by series of electric pulses. 47

STRAIN: change in shape of objects when forces are exerted on them. 166

STRAIN GAUGE: a small piece of conductive material that measures forces and moments acting on an object, by measuring their strain and converting it to an electrical signal. 166

STEREOSCOPICALLY: viewed in three dimensions. 194

SUB-ROUTINE: a support procedure into which a computer will branch on receipt of certain information from the system. 159

SYNCHRONIZED: occuring at exactly the same time periods. 146

TACHOMETER: a feedback device used for velocity measurement. 63

TEACH BOX: a device by which a human operator moves a robot through a series of points by means of a manual pressing of switches. 82

TEACH-IN teaching in which a robot is moved by means of switches. 81, 82

TEACHING: methods in which robots are instructed how to act. 81, 87, 92

TEACH PENDANT: a teach box. 82, 108

TEACH-THROUGH: teaching in which the robot is moved by hand through various steps. 81, 85, 87

THREE-FINGER GRIPPER: an end effector having three fingers for grasping. 98

TOOL CENTER POINTER (TCP): the center point of the gripper used to define the end effector location. 76

TRANSFER: conveying or moving from one place to another. 53

TRANSLATION ERROR: a lack of precision in the location of an assembled part. 134

TRANSMISSION: a device which transfers motion in indirect drive from the motor to a joint. 52

TWO-FINGER GRIPPER: a common type of end effector having two fingers to grasp objects. 96

USER PROGRAM: a program written by the robot operator. 73, 74

VACUUM GRIPPER: a device that attaches to a flat surface by creating a vacuum. 101

VERTICAL ARTICULATED ROBOT: a robot that has three revolute joints. 32, 37, 42

VISION SENSOR: a device, like a camera, that visually identifies the location of an object and inputs the locations to the controller's memory. 81, 91

WHISKERS: a type of sensor that has thin rods protruding from the end effector which signal the controller on contact with an object. 158, 163

WORK CELL: a station in which the robot is doing work. 89

WORKPIECE: a piece of material on which work is being performed. 114

WORK RATE: productivity. 15

WORLD MODELING: an advanced, experimental programming method which enables a task to be learned without being broken down into individual motions. 81, 91

WRIST JOINTS: the joints between the elbow and the end effector. 43

YAW: movement right or left. 43

ZERO VELOCITY: without any speed 48

Index

Index

Index

Index

Index